Strength, Cuttability, and Workability of Coal

Strength, Cuttability, and Workability of Coal

Nuh Bilgin

Hanifi Copur

Cemal Balci

Deniz Tumac

CRC Press
Taylor & Francis Group
Boca Raton London New York

CRC Press is an imprint of the
Taylor & Francis Group, an **informa** business

CRC Press
Taylor & Francis Group
6000 Broken Sound Parkway NW, Suite 300
Boca Raton, FL 33487-2742

First issued in paperback 2020

ISBN 13: 978-0-367-65673-7 (pbk)
ISBN 13: 978-0-8153-9550-8 (hbk)

Library of Congress Cataloging-in-Publication Data

Names: Bilgin, Nuh, author. | Copur, Hanifi, author. | Balci, Cemal, author. | Tumac, Deniz, author.
Title: Strength, cuttability, and workability of coal / Nuh Bilgin, Hanifi Copur, Cemal Balci and Deniz Tumac.
Description: First edition. | New York, NY : CRC Press/Taylor & Francis Group, 2019. | Includes bibliographical references and index.
Identifiers: LCCN 2018040933 | ISBN 9780815395508 (hardback : acid-free paper) | ISBN 9781351183109 (ebook)
Subjects: LCSH: Coal--Analysis. | Coal-mining machinery. | Coal mines and mining.
Classification: LCC TN816 .B46 2019 | DDC 662.6/2--dc23
LC record available at https://lccn.loc.gov/2018040933

Visit the Taylor & Francis Web site at
http://www.taylorandfrancis.com

and the CRC Press Web site at
http://www.crcpress.com

Contents

Foreword

My coauthors and I have agreed to dedicate this book to Dr. Ivory Evans, since he was not only a distinguished scientist, but also he had an immense influence on my scientific career. In early 1977, he was selected as my external PhD examiner, and the examination was to be held at Bretby in the Mining Research and Establishment of National Coal Board, where he was the director at that time. I was so nervous to be examined by Dr. Evans. I travelled there with my supervisor Dr. R.J. Fowell, and we had some fruitful discussions on the way, which made me calm.

Dr. Evans' first question was: "Now, we have a meeting in this building, where there are several people invited from the industry. What would you recommend to them based on the results of your research?" This was a guiding light for my academic life pointing out that research in engineering should always meet the needs of the industry.

One of the chapters of my thesis was on the effect of edge radius on disc cutter performance. This was one of the first research investigations leading to the development of constant cross section disc cutters used on full-face tunnel boring machines. Statistical analysis showed that disc cutter performance was closely related to edge radius in an exponential manner. Dr. Evans pointed out that it would be more convincing and valuable if it was supported by a theoretical work. His concern has been a guide to me all in my academic life, and I have tried to explain these two important points to all my graduate students, and I hope these two concepts will serve as a basis for the research work carried out by the readers of this book.

Nuh Bilgin
July 2018

Preface

The world currently consumes over 7,800 million tons of coal per year, which is used by a variety of sectors including power generation, iron and steel production, cement manufacturing, etc. The majority of coal is either utilized in power generation that utilizes steam coal or lignite, or iron and steel production that uses coking coal. The role of coal in power generation is set to continue. Coal currently fuels 37% of the world's electricity and is forecast to continue to supply a strategic share over the next three decades.

For a better planning of coal mining operations, it is essential to know the strength, cuttability, and workability of coal, which are interrelated to each other. In the past, a lot of research studies were published separately on each subject. The book published by Evans and Pomeroy (1966) entitled "The Strength, Fracture, and Workability of Coal" is unique in this sense of combining the related subjects. This book is not intended to replace Evans and Pomeroy's book, but contributing to the subject. The main objective of the book is to combine the research studies carried out worldwide and to produce a comprehensive book oriented to coal industry, research students, practicing engineers, and coal mine planning teams.

Authors

Nuh Bilgin graduated from the Mining Engineering Department of Istanbul Technical University (ITU). Later, he accomplished his PhD studies in the University of Newcastle Upon Tyne, United Kingdom in 1977 on "Mechanical Cutting Characteristics of Some Medium and High Strength Rocks." He joined the Mining Engineering Department of ITU in 1978. He was appointed as a full-time professor in 1989. He spent 1 year in the Colorado School of Mines, United States of America, and 1 year in University of Witwatersrand, SA as a visiting professor. He has published more than 100 papers on mechanized mining and tunneling technologies. He is currently doing consultancy to mining and tunneling companies.

Hanifi Copur graduated from the Mining Engineering Department of Istanbul Technical University and completed his PhD on rock cutting mechanics and mechanical excavation of mines and tunnels at the Colorado School of Mines in 1999. He worked as a research engineer at the Earth Mechanics Institute of the Colorado School of Mines between 1995 and 1999. He has been working as an academician in the Mining Engineering Department of the Istanbul Technical University since 1999 and currently works as the head of department and full-professor in the same department. He has written more than 100 national and international scientific and industrial research reports and 130 publications on mechanical mining and tunneling including two books. He is the founder board member and currently vice president of the Turkish Tunnelling Society.

Cemal Balci graduated from Mining Engineering Department of Istanbul Technical University. In 2004, he got PhD after completing his thesis on "Comparison of Small and Full Scale Rock Cutting Test to Select Mechanized Excavation Machines" in Mining Engineering Department of Istanbul Technical University, Mine Mechanization and Technology Division. He joined the mechanical excavation research group in Earth Mechanics Institute, Colorado School of Mines between 2001 and 2004 as a researcher. He published numerous papers on mechanized mining and tunneling technologies and worked in many national and international implementation and research projects. He is currently working in Istanbul Technical University as a full professor and board member of Turkish Tunnelling Society.

Deniz Tumac graduated from the Mining Engineering Department of Istanbul Technical University (ITU). He received his PhD on mine and tunnel mechanization in 2010. He joined the research group in the College of Earth and Mineral Sciences, Pennsylvania State University in 2010. Since 2012, he is associated with the ITU, first as an assistant professor in the Mining Engineering Department and then from 2014 as an associate professor in the same department. He has successfully completed several scientific and industrial projects and published many papers related with mechanized mining, quarrying, and tunneling. He is currently working in ITU and is a member of the Turkish Tunnelling Society and Chamber of Mining Engineers of Turkey.

1

Introduction

1.1 General

Coal has been mined in many parts of the world throughout history, and coal mining continues to be an important economic activity providing a source for almost 37% of the world's energy. Petrographic, physical, mechanical, and structural properties of coal are of great importance in coal mining engineering, where they relate to selection of mining methods, determining cuttability, workability or grindability of coal, prevention of coal bumps and bursts, dust generation, stability analysis of coal roadways, slope stability analysis, and so on.

In recent years, we have come across well-prepared books on some of the related topics of this book. These include *"Hard Rock Mass and Coal Strength"* by Arioglu and Tokgoz (2011) and *"Coal"* edited by Kural (1994). However, *"The Strength, Fracture, and Workability of Coal"* by Ivory Evans and Duncan Pomeroy published in 1966 remains a most indispensable book for researchers on the subject. Our book with the title *"Strength, Cuttability and Workability of Coal"* is not intended to replace the book by Evans and Pomeroy, and we believe that it will be never replaced since it provides a comprehensive summary of coal winning research carried out in the Mining Research Establishment of the United Kingdom's National Coal Board, carried out by highly qualified research staff. However, more than half a century has passed since publication, and the accumulated new research results from Istanbul Technical University's Mining Engineering Department and elsewhere have encouraged us to produce this book, which differs from Evans and Pomeroy's monograph with the addition of the following topics:

- World coal reserves and production outputs
- Development in coal bed methane, which is an unconventional energy source
- Breakage characteristics of lithotypes
- New studies on coal cutting mechanics with complex shaped cutters

- Trends toward mining of the thinner coal seams and cutting coal with new generation plows
- Thick seam mining and design of coal face shearers based on laboratory coal cutting tests
- Size distribution of run of mine coal excavated by a shearer and petrographic characteristics of each size fraction
- In-mine directional drilling for methane drainage and exploration in advance of mining
- New studies on porosity, permeability, physical, and mechanical characteristics of coal including point load strength, Shore scleroscope hardness, Hardgrove index
- Schmidt hammer hardness based on cleat frequency, a new classification system for coal excavatability
- Cutting, breaking, and digging the coal in opencast mining operations; continuous surface miners, bucket wheel excavators, hydraulic excavators, power shovels, and draglines.

References

Arioglu, E. and Tokgoz, N. 2011. *Design Essential of Hard Rock Mass and Coal Strength. With practical solved problems.* Evrim Publisher, Istanbul, 280 p. (in Turkish).

Evans, I. and Pomeroy, C. D. 1966. *The Strength, Fracture and Workability of Coal.* Pergamon Press, London, UK.

Kural, O. (Editor), 1994. *Coal. Resources, Properties, Utilization, Pollution.* Istanbul Technical University, Turkey, 490 p.

2

Coal in the World, Reserves, Productions, the Use of Coal in Industry and Prices

2.1 Introduction

"The industrial stomach cannot live without coal; industry is a carbonivorous animal and must have its proper food", Jules Verne
https://www.brainyquote.com/quotes/jules_verne_701515?src=t_coal.

Please see the sub-section on reserves, productions, the use of coal in the industry, and prices.

2.2 Coal Reserves in the World

Coal is a general term used for different categories of carbonized substances as defined below:

Lignite or "brown coal": This is the youngest form of coal and is used almost exclusively for electric power generation.

Sub-bituminous coal: This coal, which has spent more time underground than lignite before being recovered, is mainly used for power generation.

Bituminous coal: Older than sub-bituminous coal, this coal can be used in heat and power manufacturing applications as a coking (metallurgical, bituminous) coal, mainly for steel and aluminum production.

Anthracite (often called as bituminous coal or hard coal, or black coal): This is the oldest form of coal and used mainly for residential (house) heating.

TABLE 2.1

Coal Reserves by Region

Region	Coal Recoverable Reserves (1000 Mt)
Africa	21.5
Latin America & The Caribbean	8.3
North America	171.6
Europe	182.7
South & Central Asia	68.7
East Asia	82.2
South East Asia & Pacific	74.5
Asia together	225.4

Source: World Energy Council: Energy Resources-Coal, https://www.worldenergy.
org/data/resources/resource/coal/, May 2018.

Coal reserves by region are given in Table 2.1. From this table, it is emerging that Europe is the second largest region on recoverable coal reserves with 182.3 billion tons (Bt) after Asia, with a total number of 225.4 Bt of coal.

As published by BP (2017), at the end of 2016, within 1139.3 1000 Mt of World proved coal reserves, 816.2 Bt comes from bituminous coal and anthracite and 323.1 Bt comes from sub-bituminous and lignite.

As reported by Gupta (2013), more than 80% of the world's total proved coal reserves are located in just ten countries and reported as given below.

2.2.1 United States of America

The United States houses the world's biggest coal reserves. The nation's proved coal reserves as of December 2012 are around 237.295 Bt, being more than one quarter of the total proven coal reserves in the world.

The reserves are widely distributed across Montana, Wyoming, Illinois, western Kentucky, West Virginia, Pennsylvania, Ohio, and Texas. The Peabody Energy-operated North Antelope Rochelle coal mine in the Powder River Basin of Wyoming is the world's biggest coal mine by reserve.

2.2.2 Russia

The Russian Federation has the second biggest coal reserves in the world. The country is estimated to hold 157.01 Bt of proved coal reserves as of December 2012, accounting for about 18% of the world's total coal reserves.

Russia's major deposits include the Donetskii reserves in Moscow, the Pechora basins in Western Russia, and the Kuznetski, Kansk-Achinsk, Irkutsk, and South Yakutsk basins in Eastern Russia. More than two-thirds of the coal produced in Russia is hard coal, with Pechora and Kuznetsk basins the principal hard coal deposits. The Kansk-Achinsk Basin is known for huge deposits of sub-bituminous coal, while the Raspadskaya mine in the Kemerovo region is the largest coal mine in Russia.

2.2.3 China

China holds the third largest coal reserves in the world. Its proven coal reserves as of December 2012 are around 114.5 Bt, being about 13% of the world's total proven coal reserves. China is also the world's biggest producer and consumer of coal.

More than 70% of China's proved recoverable coal reserves are located in the north and north-west parts of the country. Shanxi and the Inner Mongolia provinces host a major chunk of accessible coal reserves in the country. The Haerwusu coal mine in Inner Mongolia is the second biggest coal mine in the world by reserve.

2.2.4 Australia

The fourth largest coal reserves in the world are found in Australia. The country is estimated to possess 76.4 Bt of proved coal reserves at the end of 2012, accounting for about 9% of the total proven coal reserves in the world.

The country's black (hard) coal reserves are mostly concentrated in New South Wales and Queensland, which together account for more than 95% of Australia's black coal output. Victoria hosts about 96% of the country's brown coal reserves. Peak Downs coal mine in the Bowen Basin of central Queensland, followed by the Mt Arthur coal mine in the Hunter Valley region of New South Wales, is the biggest Australian coal mine by reserve.

2.2.5 India

India houses the fifth biggest coal reserves in the world. The country's proved coal reserves as of December 2013 are estimated to be 60.6 Bt. India accounts for about 7% of the world's total proved coal reserves.

The major hard coal deposits of the country are located in the eastern parts of the country. The eastern states of Jharkhand, Chhattisgarh, Orissa, and West Bengal account for more than 70% of the country's coal reserves. Andhra Pradesh, Madhya Pradesh, and Maharashtra are the other significant coal producing states in India. The southern state of Tamil Nadu hosts most of the country's lignite deposits.

2.2.6 Germany

Germany has the world's sixth biggest amount, with 40.7 Bt of proven coal reserves at the end of 2012. Germany possesses about 4.7% of the world's total proved coal reserves.

The Ruhr Coal Basin in the North Rhine-Westphalia state and the Saar Basin in the south-west Germany account for more than 75% of the country's hard coal production. The Rhineland region hosts the country's largest lignite deposits. The Garzweiler open-cast coal mine in the North Rhine-Westphalia state is considered to be Europe's biggest brown coal mine.

2.2.7 Ukraine

Ukraine holds the seventh largest coal reserves in the world. The country's proved coal reserves as of December 2012 are estimated to be 33.873 Bt. Ukraine's share in the world's total proved coal reserves is 3.9%.

Most of the country's coal reserves are located in Donets Basin in Eastern Ukraine. Also known as the Donbas Coal basin, the Donets Basin is spread across three Ukrainian provinces, namely Dnipropetrovsk, Donetsk, and Luhansk. Ukraine has 149 operating coal mines, out of which 120 are state-owned and 29 are private mines. The Komsomolets Donbasu coal mine in the Donetsk Oblast is one of the biggest coal mines in the country.

2.2.8 Kazakhstan

Kazakhstan, with more than 400 coal deposits, holds the eighth largest coal reserves in the world. The country is estimated to have 33.6 Bt of proven coal reserves at the end of 2012. It accounts for approximately 3.9% of the world's total proved coal reserves.

The country's proved coal reserves are mostly concentrated in three provinces including Karaganda Oblast in Central Kazakhstan and the Pavlodar and Kostanay Oblasts in North Kazakhstan. Karaganda and Ekibastuz are the two major coal producing basins in the country. Turgay, Nizhne-Iliyskiy, and Maikuben basins are known for their lignite reserves. Bogatyr Access Komir is the biggest open cast mining company in Kazakhstan.

2.2.9 Colombia

Colombia's coal reserves put it in the ninth place. Proven coal reserves of the country as of December 2012 are estimated to be 6.746 Bt, which amount to one fifth of the proved coal reserves of Kazakhstan. It accounts for 0.71% of the world's proved coal reserves.

Colombia hosts the biggest coal reserves in South America, with reserves mostly concentrated in the Guajira peninsula. The La Guajira and Cesar departments of Colombia account for more than 85% of the country's coal reserves. Correjon is the biggest coal mine in Colombia, followed by the La Loma coal mine. Correjon is also the biggest coal mine in Latin America and the tenth biggest in the world, by reserve.

2.2.10 Canada

Canada ranks as the tenth biggest in the world, with coal reserves only slightly less than that of Colombia. The proved coal reserves of Canada as of December 2012 stood at 6.582 Bt, accounting for about 0.69% of the world's total proved coal reserves.

More than 90% of Canada's coal reserves are located in sedimentary basins in the western part of the country. Nanaimo, Bowser, Skeena, Moose River, Maritime, and Bowron River are among the principal coal bearing sedimentary basins.

Canada has a privatized coal industry. Most of the country's coal output is from open pit mines. The TransAlta-owned Highvale Mine in Alberta is the largest surface strip coal mine in the country. Campbell River in British Columbia and Grande Cache in Alberta are the only two underground coal mines operating in the country.

2.3 Production and Consumption

Coal production and consumption by year and country are given in Table 2.2 for the biggest coal producer countries (https://yearbook.enerdata.net/coal-lignite/coal-production-data.html).

TABLE 2.2

Coal Production and Consumption by Year and Country

Country	Production/Consumption (Mt)						
	2010	2011	2012	2013	2014	2015	2016
China	3316/3350	3608/3695	3678/3832	3749/3969	3640/3864	3527/3721	3210/3546
India	570/684	582/715	603/777	610/808	657/892	683/895	708/922
US	996/954	1006/910	932/807	904/837	918/831	824/722	683/661
Australia	436/133	415/128	435/127	458/116	489/111	512/114	509/114
Indonesia	325/51	405/51	446/61	489/66	485/79	488/89	459/100
Russia	300/212	297/221	331/232	328/207	334/199	348/207	359/210
South Africa	255/193	253/182	259/187	256/193	261/201	251/190	250/194
Germany	184/232	189/236	197/247	191/247	187/239	184/236	177/226
Poland	133/134	139/137	144/135	143/137	137/130	136/127	131/124
Kazakhstan	111	116	121	120	114	107	102
Colombia	74	86	89	85	89	86	94
Turkey	73/96	76/101	71/101	60/85	65/97	58/93	71/106

Source: Global Energy Statistical Yearbook 2017, https://yearbook.enerdata.net/coal-lignite/coal-production-data.html, May 2018.

As noticed from this table, there is a decrease of coal production in several counties except for India, Russia, Colombia, and Turkey. The biggest coal exporting countries are Australia, Indonesia, Russia, United States of America, South Africa, and Poland.

China, which represents 44% of the world coal and lignite production, recorded a contraction in its production for the recent years. To help supporting the domestic prices, the Chinese Government has implemented measures to reduce coal production capacities including the closure of the most inefficient mines. In addition, the country's output was also impacted by legislation from April 2016 that has cut coal mine production to a 276-day basis, down from 330 days previously. Colombia and Russia, two of the world largest coal exporters, increased their production in the second half of 2016 to supply the international market.

2.4 Use of Coal in the Industry

Coal has many uses worldwide. The most significant uses of coal are in electricity generation, steel production, cement production, and liquid fuel production, which are summarized below (https://www.worldcoal.org/coal/uses-coal):

Coal plays a vital role in electricity generation worldwide. Steam coal—also known as thermal coal—is mainly used in power generation. Currently power plants using coal fuel 37% of global electricity and, in some countries, coal fuels a higher percentage of electricity.

Coking coal—also known as metallurgical coal—is mainly used in steel production. Steel production depends on coal. Today, 74% of the steel produced uses coal. Metallurgical coal or coking coal is a vital ingredient in the steel making process. World crude steel production in 2017 was 1.6 Bt.

Coal is used as an energy source in cement production. Large amounts of energy are required to produce cement. It takes about 200 kg of coal to produce one ton of cement. It is interesting to note that 4.2 Bt of cement were produced globally in 2016. China's cement production alone reached at 2.4 Bt.

Other important users of coal include alumina refineries, paper manufacturers, and the chemical and pharmaceutical industries. Several chemical products can be produced from the by-products of coal. Refined coal tar is used in the manufacture of chemicals, such as creosote oil, naphthalene, phenol, and benzene. Ammonia gas recovered from coke ovens is used to manufacture ammonia salts, nitric acid, and agricultural fertilizers. Thousands of different products have coal or coal by-products as components: soap, aspirins, solvents, dyes, plastics, and fibers, such as rayon and nylon.

Coal is also an essential ingredient in the production of specialist products, activated carbon—used in filters for water and air purification and in kidney dialysis machines, carbon fiber—an extremely strong, but light weight reinforcement material used in construction, mountain bikes, and tennis rackets. Silicon metal is used to produce silicones and silanes, which are in turn used to make lubricants, water repellents, resins, cosmetics, hair shampoos, and toothpastes.

The pressure to provide transport infrastructure and fuels is immense. Globally, the ownership of motor vehicles has increased from around 250 million in 1970 to over one billion today. Coal-derived fuels and energy carriers, as well as coal-based electricity, can play a significant role in responding to the growing energy needs of the transport sector. Coal is also an important raw material and source of primary energy for manufacturing of materials used to build transport infrastructure, such as steel, cement, and aluminum.

Liquid fuels from coal provide a viable alternative to conventional oil products and can be used in the existing supply infrastructure. Several coal-to-liquids demonstration plants are being developed in China. Coal-to-liquids currently provide 20% of South Africa's transport needs including 7.5% of jet fuel.

2.5 Coal Prices

Coal prices by region are given in Table 2.3 (https://www.bp.com/en/global/corporate/energy-economics/statistical-review-of-world-energy/coal/coal-prices.html).

TABLE 2.3

Coal Prices by Region

Region	2011	2012	2013	2014	2015	2016
NW Europe marker $/t	121.52	92.50	81.69	75.38	56.79	59.87
US Central Spot Index $/t	87.38	72.06	71.39	69.00	53.59	53.56
Japan coking coal (cif) $/t	229.12	191.46	140.45	114.41	93.85	89.40
Japan stem coal (cif) $/t	136.21	133.61	111.16	97.65	79.47	72.97
Asian marker $/t	125.74	105.50	90.9	77.89	63.52	69.91
China Qinhuangdao spot price $/t	127.7	111.89	95.42	84.12	67.53	71.35
Japan steam spot price $/t	126.13	100.30	90.07	76.13	60.10	71.66

Source: https://www.bp.com/en/global/corporate/energy-economics/statistical-review-of-world-energy/coal/coal-prices.html, BP, Coal Prices, May 2018.

As noticed in Table 2.3, there is tremendous decrease in coal prices, which is worth searching the reasons. Coking coal in Japan market decreased from 229.2 $/t in 2011 to 89.40 $/t in 2016. United States Central Spot Index decreased from 87.38 $/t in 2011 to 53.56 $/t in 2016.

2.6 Conclusive Remarks

Coal has many uses worldwide. The most significant uses of coal are in electricity generation, steel production, cement production, and liquid fuel production. Coal plays a vital role in electricity generation worldwide. Steam coal—also known as thermal coal—is mainly used in power generation. Currently, power plants are using coal fuel 37% of global electricity and, in some countries, coal fuels a higher percentage of electricity. Coking coal—also known as metallurgical coal—is mainly used in steel production. Steel production depends on coal. Today, 74% of the steel produced uses coal. Metallurgical coal or coking coal is a vital ingredient in the steel making process. Coal is used also as an energy source in cement production. Large amounts of energy are required to produce cement. It takes about 200 kg of coal to produce one ton of cement. It is interesting to note that 4.2 Bt of cement were produced globally in 2016. Other important users of coal include alumina refineries, paper manufacturers, and the chemical and pharmaceutical industries. Liquid fuels from coal provide a viable alternative to conventional oil products and can be used in the existing supply infrastructure (https://www.worldcoal.org/coal/uses-coal).

Coal is abundant all over the world. As reported by BP (2017), at the end of 2016, world's proved coal reserves are 1,139.3 Bt. More than 80% of the world's total proved coal reserves are located in just ten countries: United States of America, Russia, China, Australia, India, Germany, Ukraine, Kazakhstan, Colombia, and Canada (Gupta 2013). However, in recent years, there is a tendency of decreasing of coal production in several counties except for India, Russia, Colombia, and Turkey. The biggest coal exporting countries are Australia, Indonesia, Russia, United States of America, South Africa, and Poland (BP 2017).

It is interesting to note that there is tremendous decrease in coal prices, which is worth searching the reasons of this negative trend (https://www.bp.com/en/global/corporate/energy-economics/statistical-review-of-world-energy/coal/coal-prices.html).

References

BP, 2017. *Statistical Review of World Energy*, 66th edition, 40 p. https://www.bp.com/en/global/.../energy.../statistical-review-of-world-energy.html.

Gupta, A. 2013. Analysis, countries with the biggest coal reserves. Online: https://www.mining-technology.com/features/feature-the-worlds-biggest-coal-reserves-by-country/.

https://www.bp.com/en/global/corporate/energy-economics/statistical-review-of-world-energy/coal/coal-prices.html, taken on May 2018, BP: Coal Prices.

https://www.brainyquote.com/quotes/jules_verne_701515?src=t_coal, taken on May 2018, Brainy Quote: Coal Quotes.

https://www.worldcoal.org/coal/uses-coal, taken on May 2018, World Coal Association: Uses of Coal.

https://www.worldenergy.org/data/resources/resource/coal/, taken on May 2018, World Energy Council: Energy Resources-Coal.

https://yearbook.enerdata.net/coal-lignite/coal-production-data.html, taken on May 2018, Global Energy Statistical Yearbook 2017.

3

Coal in Turkey, Reserves, Properties, and Characteristics

3.1 Introduction

Turkey is one of the important coal producers in the world. The reserve licenses of the coal in Turkey mainly belong to government. The coal mines are operated by public and private sectors, and more than 40,000 coal workers are employed in the coal industry. State-owned companies operating in the sector are given in below:

- Turkish Hard Coal Enterprise (TTK)
- Turkish Coal Enterprises (TKI)
- Electricity Generation Company (EUAS)
- General Directorate of Mineral Research and Exploration (MTA)

The reserves of the coal are mostly lignite and hard coal and main coal deposits formed in two different geologic times, Carboniferous and Tertiary aged, are given in Figure 3.1 (http://www.mta.gov.tr/v3.0/sayfalar/hizmetler/jeotermal-harita/images/5.jpg). Turkey's coal deposits consist almost exclusively of lignite and small amounts of hard coal. According to Oskay et al. (2013), Tertiary Neocene-aged coals are most appropriate for combustion in the thermal power plants due to the high total reserves despite the high ash yields and the low calorific values and mined in Afsin-Elbistan, Konya, and Soma Basin.

The lignite coal is commonly used for electricity generation in Turkey by covering 70%–80% of the total and about 8%–10% is consumed for heating purposes and the other industries. Lignite fired power plants are about 10,000 MW and increasing significantly in recent years operated by private sector and government. The hard coal deposits are located in the Zonguldak Basin, between Eregli and Amasra on the Black Sea coast in north-western Turkey and used for thermal power plants, steel production, industrial, and domestic heating purposes. According to Karayigit et al. (1998), Zonguldak hard coals owned by TTK are high quality and low sulfur coking coal.

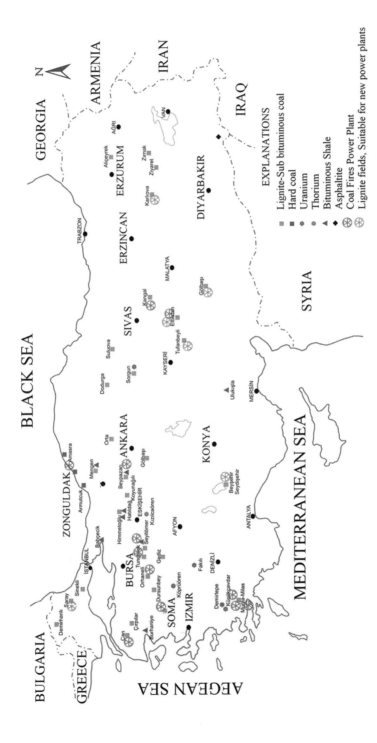

FIGURE 3.1

General coal deposits map in Turkey. (From http://www.mta.gov.tr/v3.0/sayfalar/hizmetler/jeotermal-harita/images/5.jpg.)

3.2 Reserves

Lignite coal deposits in Turkey are spread across the country, the most important one being the Afsin-Elbistan Lignite Basin of south-eastern Anatolia
and the proven reserves of about 4.6 billion tons. The Soma Basin is the
second-largest lignite mining area in Turkey and other exploited deposits are
located in Konya, Eskisehir, Mugla, Afyon, and Tekirdag. Hard coal deposits
in Turkey are generally spread in the Zonguldak Basin, between Eregli and
Amasra on the Black Sea coast. The total reserve including lignite and hard coal
is about 19 billion tons according to MTA in Turkey. Turkish lignite and hard
coal reserves are seen in Table 3.1.

TABLE 3.1

Turkish Lignite and Hard Coal Reserves by Region

Reserve Area of Turkish Coals	Reserve (ton)
Afsin-Elbistan	4,642,340,000
Afsin-Elbistan (MTA)	515,000,000
Manisa-Soma	861,450,000
Adana-Tufanbeyli	429,549,000
Adiyaman-Golbasi	57,142,000
Bingol-Karliova	88,884,000
Ankara-Beypazari	498,000,000
Afyon-Dinar-Dombayova	941,000,000
Bolu-Mengen	142,757,000
Mugla-Milas	750,214,000
Cankiri-Orta	123,165,000
Canakkale-Can	92,483,000
Kutahya-Tuncbilek	317,732,000
Kutahya-Seyitomer	198,666,000
Sivas-Kangal	202,607,000
Kutahya-Gediz	23,945,000
Tekirdag-Cerkezkoy	573,600,000
Tekirdag-Malkara	618,000,000
Tekirdag-Saray	141,175,000
Amasya-Yeniceltek	19,791,000
Yozgat-Sorgun	13,206,000
Bolu-Goynuk	43,454,000
Corum-Dodurga	38,500,000
Konya-Karapinar	1,832,000,000
Konya (Beysehir-Seydisehir)	348,000,000
Bolu (Salip-Merkez)	98,000,000
İstanbul (Silivri)	180,000,000

(Continued)

TABLE 3.1 (*Continued*)

Turkish Lignite and Hard Coal Reserves by Region

Reserve Area of Turkish Coals	Reserve (ton)
Eskisehir (Alpu)	1,453,000,000
Eskisehir (Koyunagili)	57,430,000
Edirne	90,000,000
Bursa (Keles-Orhaneli)	85,000,000
Balikesir	34,000,000
Ankara (Golbasi)	48,000,000
Other	1,928,810,000
Zonguldak Hard Coal	1,500,000,000
TOTAL	**18,978,900,000**

Source: General Directorate of Mineral Research and Exploration, http://www.mta.gov.tr/v3.0/arastirmalar/komur-arama-arastirmalari, May 2018.

The Turkish coal sector produced about 2 million tons/year of hard coal between 2011 and 2016 and about 70 million tons/year–80 million tons/year of lignite in 2011 to 2016 by private and public sectors. Turkish lignite and hard coal productions are seen in Table 3.2 by year.

3.3 Properties and Characteristics

Turkish lignite and sub-bituminous coals are ranked as low rank coals according to the origin and high ash yields and the low calorific values. Some Turkish lignites' properties in terms of moisture, ash, sulfur, volatile matter, and low heating values are compiled by Ziypak (2015) and given in Table 3.3.

The histogram and frequency analysis are carried out for the moisture, ash, sulfur, volatile matter contents, and low heating values using Table 3.3 and given in Figures 3.2 through 3.6.

The moisture, ash, sulfur, volatile matter, and low heating values of Turkish lignites vary around 13% kcal/kg–53.5% kcal/kg; 15% kcal/kg–43% kcal/kg; 0.6% kcal/kg–4.7% kcal/kg; 16% kcal/kg–34% kcal/kg; and 1,032% kcal/kg–3,150% kcal/kg, respectively.

Zonguldak's hard coal has better chemical properties and low mineral matter, sulfur, moisture, and ash contents (Toprak 2009). Some chemical properties of Zonguldak hard coal are given in Table 3.4. The calorific values of hard coals generally vary between 6,000 kcal/kg and 7,500 kcal/kg.

TABLE 3.2

Turkish Lignite and Hard Coal Production by Year

Coal Name	Sector	2011 (ton)	2012 (ton)	2013 (ton)	2014 (ton)	2015 (ton)	2016 (ton)
Asphaltite		971,366	749,830	751,263	336,852	837,112	1,567,277
Bituminous	TKI	394,277	394,277	149,828	259,508	288,185	372,028
Lignite	EUAS	31,455,812	24,297,709	16,011,459	18,987,907	10,855,125	13,267,974
	TKI	41,530,095	36,815,592	23,257,009	22,854,114	12,432,171	24,330,886
	Other public	0	0	11,245,055	1,063,927	399,816	213,501
	Private	9,389,340	16,900,893	12,810,342	23,301,062	35,043,058	40,083,687
	Total	82,375,247	78,014,194	63,323,865	66,207,010	58,730,170	77,896,048
Hard coal	TTK + Royalty	2,619,247	3,235,299	2,789,338	1,916,833	2,074,049	2,137,002
General Total		86,360,137	82,393,600	67,014,294	68,720,203	61,929,516	81,972,355

Source: Republic of Turkey Ministry of Energy and Natural Resources, http://www.enerji.gov.tr/tr-TR/Sayfalar/Komur, May 2018.

TABLE 3.3

Some Specifications of Turkish Coals

Owner of Licenses	City	District	Reserve (ton)	Moisture (%)	Ash (%)	Sulphur (%)	Volatile Matter (%)	LHV (kcal/kg)
EUAS[a]	Ankara	Beypazari	308,261,548	24–26	30	3–4	20–25	2,399–2,839
KIAS-EUAS	Ankara	Koyunagili	52,646,001	25	32	2.7	22	2,250
EUAS	K.Maras	Elbistan	4,360,163,536	50–55	17–21	1.5–4.0	19–21	1,031–1,201
PS	Sivas	Kangal	90,369,000	48–52	19–21	2.76	10–46[a]	1,207–1,494
PS	Kutahya	Seyiomer	142,009,000	32	43	1.2	22	2,080
TKI+PS	Adana	Tufanbeyli	500,000,000	41	28	2.1	24	1,298
TKI	Bingol	Karliova	130,662,000	47	24	0.6	16	1,460
TKI	Kutahya	Tuncbilek	256,527,000	15	41	1.6	25	2,560
TKI	Manisa	Soma (Eynez)	493,569,000	13	33	1.3	27	3,150
TKI	Manisa	Soma (Denis/Evciler/KO/TP)	170,457,000	18	40	1.2	20	2,080
TKI	Tekirdag	Saray	129,151,000	45	16	1.9	20	2,080
EUAS	Bursa	Keles (Harmanalan/Davultlar)	68,494,000	34–39	24	1.6–4.3	21–26	1,905–2,404
EUAS	Bursa	Orhaneli (Sagirlar/Gumuspinar)	36,578,000	24	24	2	34	2,500
PS	Mugla	Milas (Ekizkoy)	85,546,000	29	25	3.2	29	2,196
PS	Mugla	Milas (Karacahisar)	85,770,000	30	22	4.5		2,279
PS	Mugla	Milas (Husamlar)	45,058,000	30	34	1.2	28	1,607

(Continued)

TABLE 3.3 (*Continued*)

Some Specifications of Turkish Coals

Owner of Licenses	City	District	Reserve (ton)	Moisture (%)	Ash (%)	Sulphur (%)	Volatile Matter (%)	LHV (kcal/kg)
PS	Mugla	Yatagan (Eskihisar/Tinaz)	51,132,000	33–36	21–27	1.9–2.4	25–27	2,100–2,111
Ps	Mugla	Yatagan (Bayir/Turgut/Taskesik)	101,793,000	24–30	24–36	1.9–2.8	29–33	2,154–2,670
TKI+PS	Konya	Ilgin	185,000,000	45	15	3.5	25	2,200
TKI	Canakkale	Can (Cavuskoy)	73,053,000	23	25	4.2	30	3,000
EUAS	Konya	Karapinar	1,832,816,162	47	20	2.78	24	1,320
TKI	Eskisehir	Alpu	1,290,000,000	34	32	1.5	21	2,050
MTA	Tekirdag	Cerkezkoy	494,713,051	32	29	2.53	22.02	2,060
MTA	Afyon	Dinar	941,500,000	41	19–54	1.35	31.05	1,760
EUAS	K.Maras	Elbistan	515,051,800	48–50	25	2.15	22.32	950–1,115
PS	Cankiri	Orta	110,221,000	48	28	0.6	17	1,300

Source: Ziypak, M., Overview of coal in Turkey and environmental precautions Inventory Workshop 3, Research & Development, Head of Department, June 11–12, Ankara, Turkey, 2015.

[a] EUAS: Electricity Generation Company; TKI: Turkish Coal Enterprises; MTA: General Directorate of Mineral Research and Exploration; PS: Private Sector; KIAS belongs to Turkish Coal Enterprises; LHV: Low heating value.

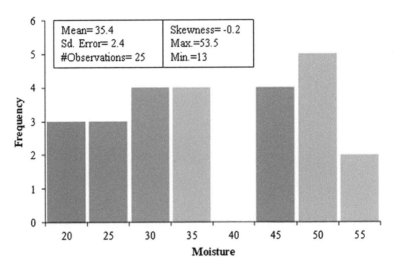

FIGURE 3.2
Frequency distribution of moisture content of Turkish lignites.

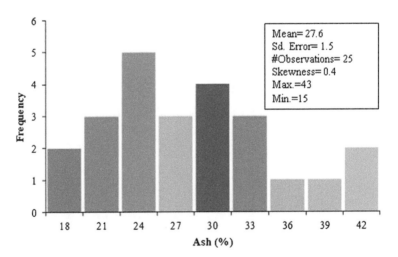

FIGURE 3.3
Frequency distribution of ash content of Turkish lignites.

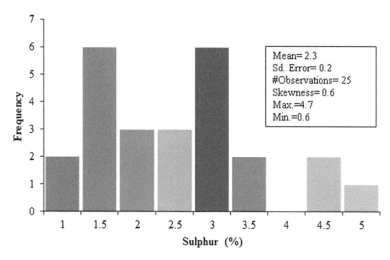

FIGURE 3.4
Frequency distribution of sulfur content of Turkish lignites.

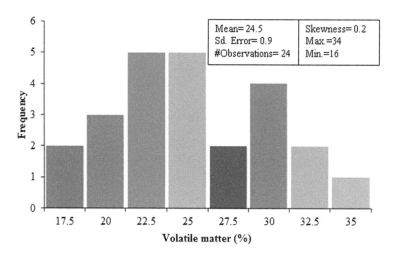

FIGURE 3.5
Frequency distribution of volatile matter of Turkish lignites.

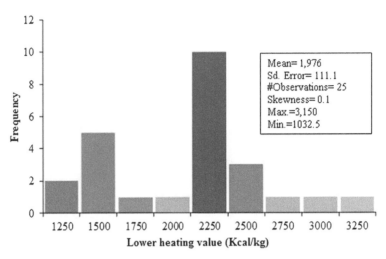

FIGURE 3.6
Frequency distribution of lower heating value of Turkish lignites.

TABLE 3.4

Some Chemical Properties of Zonguldak Hard Coal

Location	Moisture (%)	Ash (%)	Sulphur (%)	Volatile Matter (%)	LHV (kcal/kg)
Zonguldak	10	12	<1	36.02	7,500

Source: Toprak, S., *Int. J. Coal Geol.*, 78, 263–275, 2009.

Petrographic and maceral distribution of Turkish lignite coals in different regions are well collected in a book by Tuncali et al. (2002) and compiled and summarized in Table 3.5.

The histogram and frequency analysis are carried out for the hardgrove grindability index, huminite, liptinite, pyrite, clay, silt, etc., reflectance values of R_{max}, R_{min}, R_{avr}, and seam height values using values in Table 3.5 and given in Figures 3.7 through 3.16.

TABLE 3.5

Some Petrographic Properties of Turkish Lignites

Location		Seam Height (m)	Hum.	Lip.	Iner.	Pyr.	Clay, Silt etc.	R_{max}	R_{min}	R_{avg}	Std.dev.
Edirne-Demirhanli-Karayusuflu	OP	1.4	69	4	7	3	17	0.464	0.28	0.324	0.013
Edirne-Demirhanli-Hacumur	OP	1.25	77	6	6	2	9	0.419	0.202	0.303	0.058
Edirne-Meriç-Kucukdoganca	UG	1.33	79	3	4	3	11	0.381	0.28	0.324	0.02
Edirne-Kesan-Begendik	UG	0.3	79	4	3	4	10	0.428	0.295	0.369	0.002
Edirne-Kesan-Begendik	UG	0.7	79	5	4	5	7	0.494	0.195	0.315	0.065
Edirne-Kesan-Pasayigit	UG	0.9	70	3	4	4	19	0.417	0.314	0.367	0.019
Edirne-Kesan-Kucukdoganca	UG	1.02	75	5	6	6	8	0.597	0.499	0.554	0.03
Edirne-Kesan-Kucukdoganca	UG	0.5	79	6	4	3	8	0.505	0.46	0.482	0.014
Edirne-Kesan-Yenimuhacir	UG	0.8	69	7	6	4	14	0.537	0.392	0.474	0.016
Edirne-Kesan-Cobancesmesi	OP	1.4	72	5	6	4	13	0.439	0.399	0.418	0.015
Edirne-Kesan-Karacaali	OP	1.6	67	9	7	5	12	0.438	0.298	0.397	0.019
Edirne-Kesan-Cobancesmesi	UG	1.1	64	8	9	4	15	0.458	0.338	0.424	0.012
Edirne-Uzunkopru-Turkobasi	OP	0.8	62	9	6	6	17	0.493	0.365	0.455	0.013
Edirne-Uzunkopru-Gaziali	UG	0.55	62	9	7	2	20	0.476	0.335	0.422	0.018
Edirne-Uzunkopru-Harmanli	OP	2.5	67	9	8	4	12	0.542	0.459	0.501	0.017
Edirne-Uzunkopru-Cavuslu	OP	2.65	71	8	6	3	12	0.496	0.347	0.436	0.014
Edirne-Uzunkopru-Harmanli	OP	2	74	6	6	3	11	0.541	0.376	0.492	0.006

(Continued)

TABLE 3.5 (*Continued*)

Some Petrographic Properties of Turkish Lignites

Location		Seam Height (m)	Hum.	Lip.	Iner.	Pyr.	Clay, Silt etc.	R_{max}	R_{min}	R_{avg}	Std.dev.
Edirne-Uzunkopru-Harmanli	OP	3.35	70	7	8	3	12	0.472	0.37	0.432	0.017
Edirne-Uzunkopru-Karaburcak	UG	0.8	79	4	4	1	12	0.423	0.287	0.394	0.019
Edirne-Uzunkopru-Karaburcak	UG	0.4	77	6	6	3	8	0.426	0.373	0.402	0.018
Edirne-Uzunkopru-Kestanbolu	UG	1.2	72	5	5	4	14	0.444	0.36	0.416	0.02
Edirne-Uzunkopru-Kestanbolu	UG	1.15	70	6	8	5	11	0.455	0.38	0.419	0.019
Edirne-Uzunkopru-Meselik	UG	0.8	68	7	7	4	14	0.421	0.326	0.383	0.016
Edirne-Uzunkopru-Kucukdoganca	UG	1.3	63	9	9	5	14	0.393	0.254	0.376	0.017
Istanbul-Agacli	OP	1.5	81	2	6	2	9	0.447	0.262	0.366	0.02
Istanbul-Agacli	OP	1.7	73	5	6	2	14	0.424	0.359	0.393	0.012
Istanbul-Agacli	OP	2.9	63	6	5	3	23	0.413	0.271	0.395	0.013
Istanbul-Sile-Toplutepe	OP	0.8	57	7	4	5	27	0.324	0.231	0.296	0.02
Kirklareli-Pinarhisar-Tozakli	OP	1.5	75	9	8	3	5	0.416	0.297	0.396	0.017
Kirklareli-Pinarhisar	OP	2.2	80	5	6	1	8	0.425	0.334	0.393	0.014
Kirklareli-Pinarhisar-Poyrali	OP	1.4	79	4	2	3	12	0.376	0.291	0.341	0.02
Kirklareli-Pinarhisar-Akoren	OP	2.2	64	5	4	3	24	0.458	0.216	0.389	0.009
Tekirdag-Topcular	OP	1.75	74	5	5	3	13	0.452	0.259	0.391	0.015
Tekirdag-Saray-Edirkoy	OP	2.15	75	6	7	4	8	0.488	0.338	0.435	0.019

(*Continued*)

TABLE 3.5 (*Continued*)

Some Petrographic Properties of Turkish Lignites

Location		Seam Height (m)	Hum.	Lip.	Iner.	Pyr.	Clay, Silt etc.	R_{max}	R_{min}	R_{avg}	Std.dev.
Tekirdag-Malkara-Sariyar	OP	3	71	4	6	3	16	0.442	0.31	0.392	0.013
Tekirdag-Malkara-Ciftekopruler	OP	0.74	80	3	5	3	9	0.506	0.305	0.379	0.06
Tekirdag-Malkara-Haskoy	UG	0.85	72	3	7	3	15	0.375	0.201	0.312	0.014
Tekirdag-Malkara-Haskoy	OP	2.1	77	3	7	4	9	0.387	0.275	0.313	0.012
Tekirdag-Malkara-Ahmetpasa	UG	1	82	4	3	3	8	0.367	0.256	0.304	0.016
Tekirdag-Malkara-Kurtullu	OP	0.78	72	5	6	4	13	0.383	0.353	0.368	0.011
Tekirdag-Malkara-Kurtullu	UG	0.7	79	4	5	3	9	0.407	0.21	0.289	0.05
Tekirdag-Malkara-Pirinccesme	OP	1.75	87	2	2	3	6	0.42	0.311	0.369	0.02
Tekirdag-Malkara-Davuteli	OP	1.02	82	3	4	5	6	0.472	0.37	0.418	0.019
Tekirdag-Malkara-Batkin	OP	1.6	77	3	5	2	13	0.373	0.332	0.348	0.014
Tekirdag-Malkara-Karamurat	UG	2	76	5	7	2	10	0.383	0.249	0.342	0.02
Tekirdag-Malkara-Kirikali	UG	1.6	79	4	4	3	10	0.454	0.239	0.315	0.053
Tekirdag-Malkara-Ortadere	UG	0.87	45	3	4	4	44	0.393	0.325	0.364	0.012
Tekirdag-Malkara-Baglarici	UG	0.85	81	4	4	3	8	0.461	0.235	0.351	0.053
Tekirdag-Malkara-Davuteli	OP	0.3	76	5	5	3	11	0.473	0.398	0.44	0.025
Tekirdag-Malkara-Davuteli	UG	0.8	79	5	4	4	8	0.463	0.392	0.427	0.02
Tekirdag-Malkara-Davuteli	UG	0.8	73	9	7	3	8	0.482	0.386	0.435	0.013
Balikesir-Dursunbey-Odakoy	OP	5.1	80	5	3	2	10	0.392	0.26	0.349	0.014
Balikesir-Dursunbey-Cakirca	OP	4.2	74	7	4	2	13	0.324	0.244	0.289	0.019

(Continued)

TABLE 3.5 (*Continued*)

Some Petrographic Properties of Turkish Lignites

Location		Seam Height (m)	Hum.	Lip.	Iner.	Pyr.	Clay, Silt etc.	R_{max}	R_{min}	R_{avg}	Std.dev.
Balikesir-Dursunbey-Hamzacik	OP	4.15	90	5	1	2	2	0.374	0.27	0.342	0.016
Balikesir-Dursunbey-Secdere	OP	3.15	53	6	7	4	30	0.402	0.318	0.355	0.014
Balikesir-Balya-Mancilik	OP	8.8	61	6	4	6	23	0.406	0.362	0.325	0.032
Balikesir-Sindirgi-Kinik	UG	0.8	60	3	6	5	26	0.515	0.391	0.493	0.02
Balikesir-Balya-Bengiler	UG	8.5	59	5	6	7	23	0.465	0.333	0.414	0.032
Balikesir-Gonen-Catak	OP	3.55	62	4	4	4	26	0.432	0.357	0.405	0.016
Balikesir-Gonen-Sebepli	OP	8.6	59	5	3	3	30	0.728	0.475	0.614	0.03
Balikesir-Gonen-Tutuncu	OP	6.3	74	5	4	3	14	0.457	0.322	0.399	0.016
Balikesir-Balya-Derekoy	OP	1.5	66	6	5	4	19	0.439	0.313	0.382	0.012
Balikesir-Gonen-Saroluk	OP	9.5	67	5	8	6	14	0.383	0.304	0.344	0.02
Bursa-Orhaneli-Burmu	OP	7.1	79	5	5	4	8	0.423	0.238	0.392	0.009
Bursa-Harmancik-Kozluca	OP	1.5	66	6	6	4	18	0.432	0.321	0.378	0.018
Bursa-Keles-Harmanalan	OP	9.1	76	5	8	2	9	0.403	0.246	0.31	0.014
Bursa-Mustafakemalpasa-Caltilibuk	OP	3.2	55	3	4	5	33	0.488	0.392	0.461	0.02
Bursa-Mustafakemalpasa-Karacalar	OP	6.4	68	4	4	4	20	0.419	0.312	0.39	0.011
Bursa-Mustafakemalpasa-Alpagut	UG	10.5	63	4	4	5	24	0.405	0.328	0.394	0.012
Canakkale-Yenice-Cirpilar	OP	10.5	73	3	2	4	18	0.495	0.392	0.458	0.011
Canakkale-Yenice-Kalkim-Orencik	OP	1.6	76	5	5	4	10	0.526	0.459	0.492	0.014
Canakkale-Can-Etili	OP	1.3	76	3	2	5	14	0.376	0.284	0.345	0.014

(*Continued*)

TABLE 3.5 (*Continued*)

Some Petrographic Properties of Turkish Lignites

Location		Seam Height (m)	Hum.	Lip.	Iner.	Pyr.	Clay, Silt etc.	R_{max}	R_{min}	R_{avg}	Std.dev.
Canakkale-Can-Yeniceri	OP	3	72	6	4	5	13	0.432	0.361	0.394	0.012
Canakkale-Can-Durali	OP	16	65	7	5	3	20	0.374	0.312	0.351	0.015
Canakkale-Can-Comakli	OP	16.85	61	6	5	4	24	0.451	0.333	0.389	0.018
Yalova-Safran	OP	6.15	61	7	5	4	23	0.486	0.363	0.433	0.014
Afyon-Sincanli-Karacaoren	UG	1.3	67	7	5	6	15	0.372	0.261	0.312	0.02
Afyon-Suhut-Isali	UG	5.9	66	7	6	5	16	0.371	0.202	0.305	0.017
Aydin-Bozdogan-Korteke	UG	1.25	72	8	5	4	11	0.44	0.327	0.406	0.013
Aydin-Incirliova-Ikizdere	UG	1.4	66	4	3	6	21	0.431	0.358	0.409	0.013
Aydin-Nazilli-Haskoy	UG	4.5	73	7	5	4	11	0.412	0.297	0.381	0.012
Aydin-Kocarli-Mersinbeleni	UG	1.5	61	6	4	5	24	0.412	0.273	0.369	0.014
Aydin-Kosk-Bascayir	UG	0.6	67	5	8	2	18	0.489	0.397	0.459	0.019
Aydin-Kosk-Kizilcayer	UG	0.5	68	7	5	6	14	0.565	0.456	0.507	0.019
Aydin-Dalama-Kuloglu	UG	0.9	66	9	5	4	16	0.428	0.379	0.402	0.015
Aydin-Kuyucak-Saricaova	UG	0.75	78	6	7	1	8	0.415	0.196	0.387	0.014
Aydin-Sahineli	UG	7.26	82	5	7	1	5	0.463	0.393	0.441	0.02
Aydin-Soke	UG	3.35	72	8	10	4	6	0.398	0.228	0.308	0.022
Denizli-Kale-Demirciler	UG	1.15	60	5	3	3	29	0.514	0.302	0.479	0.021
Denizli-Sazak-Bostanyeri	UG	2.1	61	6	7	4	22	0.412	0.234	0.398	0.011
Denizli-Kale-Kurbalik	UG	3.65	76	6	4	4	10	0.453	0.194	0.29	0.06
Denizli-Civril-Uctepeler	UG	1.8	71	4	5	6	14	0.431	0.356	0.405	0.011
Denizli-Civril-Uctepeler	OP	1.8	70	6	6	5	13	0.485	0.406	0.456	0.02
Denizli-Civril-Tokca	UG	0.8	78	6	5	4	7	0.469	0.361	0.419	0.014
Denizli-Saraykoy-Kabaagac	UG	1	62	8	8	2	20	0.429	0.317	0.392	0.02
Denizli-Cardak-Hayrettin	UG	1.45	42	3	3	4	48	0.422	0.301	0.372	0.016

(Continued)

TABLE 3.5 (*Continued*)

Some Petrographic Properties of Turkish Lignites

Location		Seam Height (m)	Hum.	Lip.	Iner.	Pyr.	Clay, Silt etc.	R_{max}	R_{min}	R_{avg}	Std.dev.
Denizli-Cameli-Karabayir	UG	0.8	75	8	6	5	6	0.486	0.353	0.419	0.012
Denizli-Cameli-Cumaalani	UG	1	66	6	5	5	18	0.465	0.311	0.398	0.015
İzmir-Cumaovasi-Bahcecik	UG	1.25	70	4	2	6	18	0.91	0.764	0.801	0.02
Kutahya-Gediz-Gokler	OP	2	66	5	5	6	18	0.489	0.333	0.428	0.02
Kutahya-Gediz-Sazkoy	UG	6.1	61	3	2	4	30	0.476	0.359	0.405	0.019
Kutahya-Seyitomer-Siroren	OP	4.25	67	5	5	5	18	0.391	0.314	0.371	0.028
Kutahya-Seyitomer-Darica	OP	3.05	61	5	5	6	23	0.365	0.218	0.335	0.016
Kutahya-Seyitomer-Aslanli	OP	13	76	7	4	3	10	0.405	0.245	0.351	0.015
Kutahya-Seyitomer-Kizik	OP	5.35	76	6	7	3	8	0.429	0.352	0.411	0.012
Kutahya-Seyitomer-Gulbektepe	OP	12.7	64	5	6	5	20	0.438	0.351	0.405	0.015
Kutahya-Seyitomer-Cobankoy	OP	1	76	5	5	4	10	0.432	0.304	0.386	0.011
Kutahya-Seyitomer-Kepez	OP	1.25	59	5	4	4	28	0.378	0.304	0.356	0.012
Kutahya-Seyitomer-Isakoy	OP	0.9	78	4	4	5	9	0.427	0.328	0.408	0.013
Kutahya-Tavsanli-Opanoz	OP	0.65	73	6	6	5	10	0.405	0.323	0.391	0.016
Kutahya-Tavsanli-Emirler	OP	1.45	79	5	5	3	8	0.503	0.412	0.481	0.014
Kutahya-Tavsanli-Leylekkiran	OP	1.75	66	4	4	5	21	0.438	0.359	0.399	0.021
Kutahya-Tavsanli-Yenikoy	OP	1.05	72	10	4	5	9	0.42	0.264	0.368	0.019
Kutahya-Tavsanli-Dereli	OP	1	70	8	4	2	16	0.517	0.384	0.466	0.023
Kutahya-Tavsanli-Degirmisaz	OP	5.8	70	6	3	5	16	0.782	0.644	0.713	0.013
Kutahya-Tavsanli	OP	3.9	65	8	5	6	16	0.504	0.404	0.455	0.031

(Continued)

TABLE 3.5 (*Continued*)

Some Petrographic Properties of Turkish Lignites

Location		Seam Height (m)	Hum.	Lip.	Iner.	Pyr.	Clay, Silt etc.	R_{max}	R_{min}	R_{avg}	Std.dev.
Kutahya-Hisarcik-Catak	UG	1	61	4	9	3	23	0.439	0.377	0.402	0.026
Kutahya-Tuncbilek-Hamitabat	OP	7.5	72	5	4	5	14	0.506	0.375	0.479	0.019
Kutahya-Tuncbilek-Omerler	UG	1.3	71	5	5	4	15	0.547	0.393	0.497	0.011
Kutahya-Tuncbilek-Karakaya	OP	1.5	66	6	7	5	16	0.385	0.338	0.365	0.015
Kutahya-Tuncbilek-GLI(D12)	OP	5.25	58	8	7	4	23	0.399	0.316	0.365	0.024
Kutahya-Tuncbilek-GLI(6C)	OP	8.55	72	5	6	3	14	0.57	0.315	0.505	0.016
Kutahya-Tuncbilek-GLI(BY11)	OP	8.15	70	7	5	3	15	0.656	0.451	0.576	0.018
Kutahya-Altintas-Oysu	UG	1.2	31	4	4	8	53	0.55	0.393	0.497	0.016
Manisa-Soma-Eynez	UG	24.5	73	5	5	1	16	0.518	0.379	0.472	0.012
Manisa-Soma-Isiklar	UG	27.88	87	6	2	2	3	0.54	0.374	0.479	0.019
Manisa-Akhisar-Dagdere	UG	2.25	69	10	5	4	12	0.365	0.214	0.306	0.011
Manisa-Akhisar-Kavakalan	UG	0.7	72	7	8	4	9	0.494	0.402	0.447	0.02
Manisa-Soma-Kisrakdere	OP	24.56	79	8	5	1	7	0.434	0.211	0.379	0.014
Manisa-Soma-Sarikaya	OP	10.23	76	6	3	2	13	0.436	0.266	0.332	0.019
Manisa-Soma-Elmali	OP	17.85	62	5	4	3	26	0.435	0.263	0.364	0.012
Manisa-Soma-Darkale	UG	11.45	77	5	4	3	11	0.406	0.267	0.353	0.01
Manisa-Kirkagac-Gelenbe	UG	0.52	42	3	2	3	50	0.384	0.244	0.342	0.015
Manisa-Gordes-Citak	UG	2.8	72	7	4	2	15	0.443	0.237	0.345	0.014
Manisa-Gordes-Kalemoglu	UG	1	64	6	3	3	24	0.414	0.223	0.385	0.009
Manisa-Soma-Tarhala	UG	8.5	71	6	3	3	17	0.421	0.336	0.377	0.012

(*Continued*)

TABLE 3.5 (*Continued*)

Some Petrographic Properties of Turkish Lignites

Location		Seam Height (m)	Hum.	Lip.	Iner.	Pyr.	Clay, Silt etc.	R_{max}	R_{min}	R_{avg}	Std.dev.
Manisa-Soma-Denis(I)	OP	15.35	80	6	4	2	8	0.411	0.198	0.377	0.019
Manisa-Kula-Pabuclu	OP	0.35	56	5	4	6	29	0.378	0.218	0.336	0.014
Manisa-Soma-Denis(II)	OP	11.9	65	3	3	3	26	0.449	0.317	0.405	0.013
Mugla-Milas-Cakiralan	OP	1.2	57	7	7	3	26	0.506	0.396	0.492	0.012
Mugla-Milas-Alatepe	OP	1.76	78	4	8	3	7	0.507	0.31	0.474	0.02
Mugla-Goktepe-Berdik	UG	1.98	63	9	7	3	18	0.379	0.225	0.33	0.014
Mugla-Yerkesik-Kultak	UG	1.1	73	7	4	4	12	0.502	0.289	0.436	0.006
Mugla-Yatagan-Bagyaka	OP	12.3	58	8	6	4	24	0.38	0.214	0.282	0.012
Mugla-Milas-Ekizkoy	OP	8.95	72	5	3	2	18	0.334	0.278	0.306	0.017
Mugla-Milas-Sekkoy	OP	14.2	55	5	7	5	28	0.312	0.168	0.234	0.015
Mugla-Milas-Husamlar	OP	6.45	52	6	4	3	35	0.354	0.17	0.314	0.011
Mugla-Yatagan-Eskihisar	OP	11.95	58	6	3	4	29	0.349	0.203	0.298	0.02
Mugla-Yatagan-Bayir (Merdivenli)	OP	6.94	57	9	6	6	22	0.368	0.228	0.328	0.018
Mugla-Yatagan-Tinaz	OP	13.32	53	7	4	4	32	0.319	0.196	0.274	0.013
Usak-Ilyasli	UG	1	71	5	5	3	16	0.407	0.303	0.376	0.013
Amasya-Sultuova-Eskiceltek	OP	5	75	5	7	3	10	0.517	0.42	0.481	0.031
Bolu-Mengen-Merkesler (Izler)	UG	3.3	69	8	6	5	12	0.488	0.407	0.468	0.02
Bolu-Mengen-Gokcesu (Salipazari)	UG	8.55	73	6	6	2	13	0.458	0.317	0.397	0.019
Bolu-Mengen-Gokcesu (Salipazari)	UG	2.95	72	7	8	4	9	0.42	0.353	0.392	0.019
Bolu-Goynuk-Himmetoglu	OP	8	72	6	6	4	12	0.423	0.297	0.38	0.014

(*Continued*)

TABLE 3.5 (*Continued*)

Some Petrographic Properties of Turkish Lignites

Location		Seam Height (m)	Hum.	Lip.	Iner.	Pyr.	Clay, Silt etc.	R_{max}	R_{min}	R_{avg}	Std.dev.
Corum-Alpagut-Dodurga	OP	33.15	72	5	9	4	10	0.53	0.427	0.478	0.024
Corum-Bayat-Karakaya	UG	2.5	74	6	5	4	11	0.513	0.415	0.476	0.023
Kastamonu-Tosya-Aspiras	OP	3.1	67	7	7	5	14	0.38	0.296	0.367	0.012
Kastamonu-Tosya-Karhin	OP	0.8	65	9	7	6	13	0.385	0.274	0.349	0.015
Tokat-Artova	UG	3.8	67	7	5	4	17	0.467	0.331	0.418	0.029
Ankara-Beypazari-Cayirhan	UG	3	67	7	6	4	16	0.365	0.282	0.312	0.016
Cankiri-Ilgaz-Ilisilik	UG	1.6	65	2	4	10	19	0.468	0.343	0.411	0.017
Eskisehir-Mihalliccik	UG	1.18	73	2	2	5	18	0.44	0.307	0.395	0.014
Konya-Ilgin-Haremi	OP	12.5	76	6	6	3	9	0.476	0.391	0.444	0.016
Konya-Seydisehir-Bayavsar	OP	2.5	72	7	7	5	9	0.429	0.368	0.396	0.019
Nevsehir-Gulsehir-Dadagi (Alemli)	UG	1.22	70	7	8	3	12	0.559	0.494	0.532	0.017
Sivas-Kangal-Kalburcayiri	OP	8.6	43	6	7	6	38	0.334	0.17	0.268	0.025
Sivas-Gemerek	OP	1.6	79	6	7	3	5	0.452	0.249	0.385	0.017
Yozgat-Sorgun-Yeniceltek	OP	9	78	6	4	3	9	0.454	0.313	0.416	0.011
Adana-Tufanbeyli-Yamanlar	OP	18.3	53	5	5	6	31	0.373	0.229	0.298	0.015
Burdur-Elmali-Pirmaz	OP	1.4	76	4	5	5	10	0.44	0.315	0.399	0.02
Isparta-Yalvac-Yukarikasikara	OP	0.7	50	5	4	5	36	0.314	0.107	0.193	0.016
Karaman-Ermenek-Tepebasi	UG	4.8	58	8	18	5	11	0.426	0.351	0.392	0.02
Karaman-Ermenek-Cenne	UG	4.55	68	9	8	4	11	0.527	0.441	0.487	0.02
Karaman-Ermenek-Canakci	UG	13.5	64	9	10	4	13	0.423	0.3	0.389	0.019
Karaman-Ermenek-Canakci	OP	7	73	7	8	3	9	0.412	0.312	0.382	0.02
Icel-Namrun-Camliyayla	OP	0.4	67	8	10	3	12	0.517	0.366	0.474	0.015

(*Continued*)

TABLE 3.5 (*Continued*)

Some Petrographic Properties of Turkish Lignites

Location		Seam Height (m)	Hum.	Lip.	Iner.	Pyr.	Clay, Silt etc.	R_{max}	R_{min}	R_{avg}	Std.dev.
Adiyaman-Golbasi	OP	29.5	58	5	7	5	21	0.344	0.161	0.302	0.019
Agri-Eleskirt-Yigiltas	OP	2.9	48	7	7	5	33	0.355	0.199	0.321	0.014
Erzurum-Horasan-Aliceyrek	OP	4.45	57	6	5	4	28	0.385	0.241	0.341	0.015
Erzurum-Oltu-Sutkans	UG	2	78	7	4	3	8	0.579	0.432	0.5	0.019
Erzurum-Ispir-Karahan	UG	2.75	56	7	5	6	26	0.381	0.28	0.335	0.014
Erzurum-Oltu-Balkaya	UG	3.25	71	7	5	4	13	0.617	0.52	0.569	0.02
Erzurum-Askale-Kukurtlu	UG	3.85	70	4	6	6	14	0.603	0.503	0.547	0.019
Kahramanmaras-Elbistan-Kislakoy	OP	30	71	3	9	1	16	0.386	0.26	0.328	0.011
Kahramanmaras-Elbistan-Kislakoy(a)	OP	36	68	3	6	2	21	0.39	0.298	0.348	0.014

Source: Tuncali, E. et al., *Chemical and Technological Properties of Turkish Tertiary Coals*, MTA Publication, Ankara, Turkey, 402 p, 2002.

Note: UG: underground; OP: open pit; Hum: Huminite; Lip: Liptinite; Iner: Inertinite; Pyr: Pyrite; R_{max}: Maximum reflectance; R_{min}: Minimum reflectance; R_{avg}: Average reflectance.

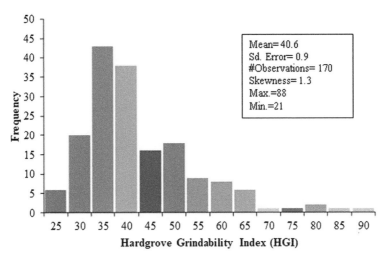

FIGURE 3.7
Frequency distribution of hardgrove grindability index (HGI) of Turkish lignites.

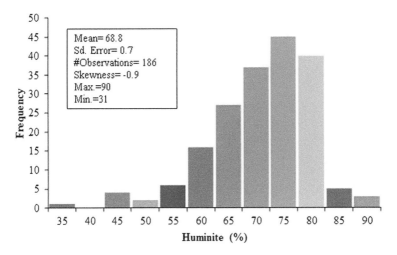

FIGURE 3.8
Frequency distribution of huminite of Turkish lignites.

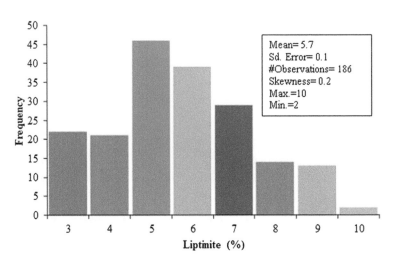

FIGURE 3.9
Frequency distribution of liptinite of Turkish lignites.

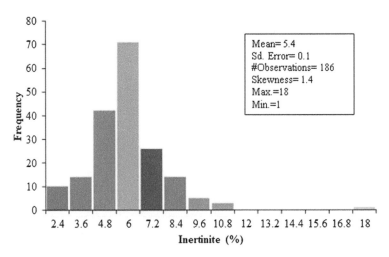

FIGURE 3.10
Frequency distribution of inertinite of Turkish lignites.

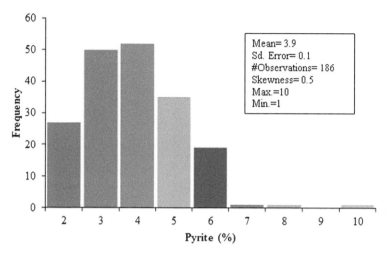

FIGURE 3.11
Frequency distribution of pyrite of Turkish lignites.

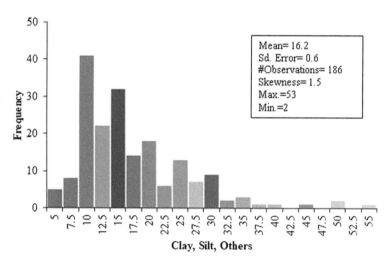

FIGURE 3.12
Frequency distribution of clay, silt, others of Turkish lignites.

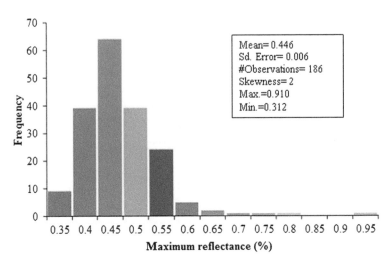

FIGURE 3.13
Frequency distribution of maximum reflectance (R_{max}) of Turkish lignites.

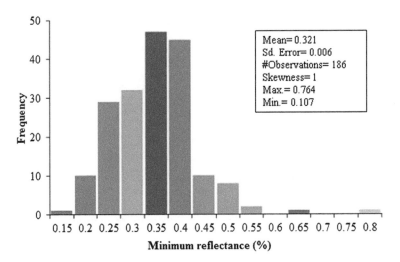

FIGURE 3.14
Frequency distribution of minimum reflectance (R_{min}) of Turkish lignites.

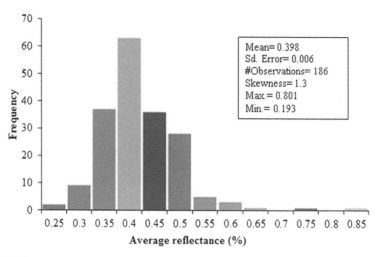

FIGURE 3.15
Frequency distribution of average reflectance (R_{avg}) of Turkish lignites.

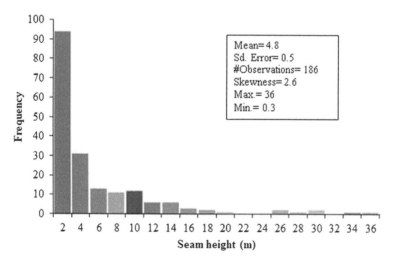

FIGURE 3.16
Frequency distribution of seam height of Turkish lignites.

The hardgrove grindability index, huminite, liptinite, pyrite, clay, silt etc., R_{max}, R_{min}, R_{avg}, and seam height values of Turkish lignites vary around 21% m–88% m, 31% m–90% m, 2% m–10% m, 1% m–18% m, 1% m–10% m, 2% m–53% m, 0.312% m–0.910% m, 0.107% m–0.764% m, 0.193% m–0.801% m, and 0.3% m–36% m, respectively.

Petrographic and maceral properties of Zonguldak's hard coal are given in Table 3.6 (Toprak 2009). Maceral group distribution of the major coals in Turkey is given in Figure 3.17 (Toprak 2009).

As seen in Figure 3.17, huminite (vitrinite for Zonguldak region) is the most dominant maceral group. Small amounts (mostly less than 7%) of liptinite and inertinite macerals are observed in the coals. The coals seem to comprise

TABLE 3.6

Maceral Distribution of Turkish Hard Coal

Location	Maximum Reflectance (%)	Vitrinite (%)	Liptinite (%)	Inertinite (%)	Pyrite (%)	Inorganic (%)
Caydamar Seam	1.02	81	5	11	1	2
Karadon Fm. Seam	1.05	82	2	14	1	1
Sulu Seam	1.04	75	8	16	1	0

Source: Toprak, S., *Int. J. Coal Geol.*, 78, 263–275, 2009.

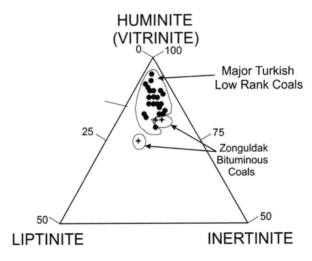

FIGURE 3.17
Maceral group distribution of the major coals in Turkey. (From Toprak, S., *Int. J. Coal Geol.*, 78, 263–275, 2009.)

considerably high amounts of clay minerals, quartz, and calcite. Pyrite and other sulfur containing minerals such as marcasite, melanterite, bilinite, and jarosite are also important in the content of the coals. Framboidal pyrite is dominant (Toprak 1996, 2009).

3.4 Conclusive Remarks

Turkish coals geologically were formed during the Miocene, Eocene, Jurassic, and Pliocene ages. Turkish lignite and sub-bituminous coal's reserve, some characteristics, chemical, and petrographic properties are briefly described and explained in this chapter. The total reserve including lignite and hard coal is about 19 billion tons according to MTA in Turkey. Some Turkish lignites' and sub-bituminous coals' properties in terms of moisture, ash, sulfur, volatile matter, and low heating values are given and frequency analysis is carried out for each parameter. The moisture, ash, sulfur, volatile matter, and low heating values of Turkish lignites vary around 13% kcal/kg–53.5% kcal/kg; 15% kcal/kg–43% kcal/kg; 0.6% kcal/kg–4.7% kcal/kg; 16% kcal/kg–34% kcal/kg; and 1,032% kcal/kg–3,150% kcal/kg, respectively. The calorific values of Zonguldak's hard coal generally vary around 6,000 kcal/kg–7,500 kcal/kg. The hardgrove grindability index, huminite, liptinite, pyrite, clay, silt etc., R_{max}, R_{min}, R_{avg}, and seam height values of Turkish lignites vary around 21% m–88% m, 31% m–90% m, 2% m–10%m, 1% m–18% m, 1% m–10% m, 2% m–53% m, 0.312% m–0.910% m, 0.107% m–0.764% m, 0.193% m–0.801%m, and 0.3% m–36% m, respectively. The hardgrove grindability index of Zonguldak Hard Coal Basin is changing between 46.96 and 96.85, according to Su et al. (2004).

References

http://www.mta.gov.tr/v3.0/sayfalar/hizmetler/jeotermal-harita/images/5.jpg taken on May 2018, General Directorate of Mineral Research and Exploration.

http://www.mta.gov.tr/v3.0/arastirmalar/komur-arama-arastirmalari taken on May 2018, General Directorate of Mineral Research and Exploration.

http://www.enerji.gov.tr/tr-TR/Sayfalar/Komur taken on May 2018, Republic of Turkey Ministry of Energy and Natural Resources.

Karayigit, A. I., Gayer, R. A., and Demirel, I. H. 1998. Coal rank and petrography of Upper Carboniferous seams in the Amasra coalfield, Turkey. *International Journal of Coal Geology*, 36:277–294.

Oskay, R. G., Inaner, H., Karayigit, A., and Christanis, K. 2013. Coal deposits of Turkey: Properties and importance on energy demand. *Bulletin of the Geological Society of Greece*, 47(4):2111–2120.

Su, O., Akcin, N., and Toroglu, I. 2004. The relationships between grindability and strength index properties of coal. *Proceedings of the 14th Turkey Coal Congress*, June 2–4, 2004, Zonguldak, Turkey (in Turkish).

Toprak, S. 1996. Determination of depositional environments and properties of coals located in the vicinity of Alpagut–Dodurga (Osmancik–Çorum) region. PhD. thesis, University of Hacettepe, Ankara, Turkey (in Turkish).

Toprak, S. 2009. Petrographic properties of major coal seams in Turkey and their formation. *International Journal of Coal Geology*, 78(4):263–275.

Tuncali, E., Ciftci, B., Yavuz, N., Toprak, S., Koker, A., Gencer, Z., Aycik, H., and Sahin, N. 2002. *Chemical and Technological Properties of Turkish Tertiary Coals.* MTA Publication, Ankara, Turkey, 402 p.

Ziypak, M. 2015. Overview of coal in Turkey and environmental precautions Inventory Workshop 3, Research & Development, Head of Department. June 11–12, 2015, Ankara, Turkey. Online: https://byt.cevre.gov.tr/Pictures/Files/Editor/document/June%202015/EN/1.7%20-%20Overview%20of%20coal%20in%20Turkey%20and%20environmental%20precautions%20-%20EN.pdf.

4

Coal Petrology, Classification, Hardness, and Breakage Characteristics of Lithotypes

4.1 Introduction

Coal is an organic sedimentary rock containing various amounts of carbon, hydrogen, nitrogen, oxygen, sulfur, and trace elements, including minerals, as well as methane, which is found within the pore systems of the coal. It is a solid, brittle, combustible, carbonaceous rock formed by the decomposition and alteration of plants by compaction, temperature, and pressure and varies in color from brown to black and is usually stratified as a coal seam. However, there are actually two main types of coal: "thermal" coal, which is mostly used for power generation and "metallurgical" (coking) coal, which is mostly used for steel production. Thermal coal is more abundant, has lower carbon content, and is higher in moisture than metallurgical coal.

Coal forms from the accumulation of plant debris, usually in a swamp environment, which have undergone a tremendous transformation while still keeping their original features. Evidence of the types of plants that contributed to carboniferous deposits can occasionally be found in the coal as fossil imprints, as well as in the shale and sandstone sediments that overlie coal deposits. The degree of change undergone by a coal as it matures from peat to anthracite is known as coalification or rank of coal, which have important effects on its physical and chemical properties. The ranks of coals, from those with the least carbon to those with the most carbon, are lignite, sub-bituminous, bituminous, and anthracite. Types of plants from which the coal originated, depths of burial, temperatures and pressures at those depths, and length of time the coal has been forming in the deposit determine the quality of each coal deposit.

Coalification or metamorphosis of coal is defined as gradual changes in the physical and chemical properties of coal in response to temperature and time. The coal changes from peat through lignite and bituminous coal to anthracite. With extreme metamorphism, anthracite can change to graphite. The rank of coal is the stage that the coal has reached on the coalification path.

The changes, with increasing rank, include an increase in carbon content and decreases in moisture content and volatile matter. Water and carbon dioxide are produced during coalification of lower ranks of coal.

Coal is a non-homogeneous material, which is composed of a number of microscopic organic and inorganic constituents. The organic constituents are called macerals and the inorganic constituents minerals. Macerals and minerals occur together in various proportions, forming macroscopic and microscopic bands, which are called lithotypes and microlithotypes (Stach et al. 1982). The main objective of this chapter is to discuss the characteristics of these lithotypes in relation to contribution to fracture and cuttability mechanism of coal.

4.2 Breakages and Strength Characteristics of Lithotypes

A series of layers found within the coal seam forms the macroscopic bands known as lithotypes, for example, vitrain, durain, clarain, and fusain. Each lithotype has a particular mechanical strength, which causes structural heterogeneity in banded coal seams, and these are related directly to the strength of the seam (Falcon and Falcon 1987). The frequency of fissures is highest in vitrinites and the layers of vitrain. This is the reason why in run-of-mine coal, durites, and most of the trimacerites are concentrated in the coarser size fraction (Stach et al. 1982). During mining, it has been shown that the energy required to break vitrain, clarain, and durain was two, three, and seven times higher, respectively, than that of the required to break fusain. It has also been found that the amount of power required to mine tough durain coal by a continuous miner was as much as 40% higher than that of the required for a friable clarain coal (Falcon 1978a, 1978b).

After ASTM D 121-09 and Stach et al. (1982), lithotypes are defined as follows:

Lithotype: It is any of the constituents of banded coal: vitrain, fusain, clarain, durain, or attrital coal or a specific mixture of two or more of these.

Attrital coal: It is the ground mass or matrix of banded coal in which vitrain and, commonly, fusain layers as well, are embedded or enclosed. Layers in banded coal, often referred to as bands, are commonly 1 mm to 30 mm thick. Attrital coal in banded coal is highly varied in composition and appearance, its luster varying from a brilliance nearly equal to that of the associated vitrain to nearly as dull as fusain, it exhibits striated, granulose, or rough texture. In a few cases, relatively thick layers of such attrital coal are found that contain no interbedded vitrain. Non-banded coal also is attrital coal, but is not usually referred to as such. In contrast to the coarser and more variable texture of attrital coal in banded coal, non-banded coal is notably uniform and fine in texture, being derived from size-sorted plant debris. The luster of attrital coal, which ranges from bright (but less than that of associated vitrain) to dull, is commonly used to describe and characterize attrital coal. As an alternative, some petrographers sub-divide attrital coal into clarain and durain. Clarain has bright luster and silky texture, being finely striated parallel to the coal bedding. Durain has dull luster and sometimes is referred to as dull coal. Similarly, coal consisting of vitrain or clarain or a mixture of the two is sometimes referred to as bright coal.

Vitrain: It is shiny and black, thicker than 0.5 mm, of sub-bituminous, and higher rank banded coal. It is attributed to the coalification of relatively large fragments of wood and bark, may range up to about 30 mm thick in eastern North American coals, but may be much thicker in the younger western deposits. Vitrain is commonly traversed by many fine cracks oriented normal to the banding. In lignite, the remains of woody material lack the shiny luster of vitrain in the higher rank coals and may instead be called previtrain. It is differentiated from attrital bands of lignite by its smoother texture, often showing the grain of wood. Previtrain may be several inches thick.

Clarain: It designates very finely stratified coal layers with a thickness between 3 mm and 10 mm, having a luster between that of vitrain and durain. It consists of alternating thin layers of vitrain, durain, and sometimes of fusain (Stach et al. 1982).

Durain: It can be black or grey. It is always dull. Black durain can have a grassy luster. Their fracture surfaces are rough. Durain has a thickness changing between 3 mm and 10 mm. Thinner dull bands are recorded as clarain.

Fusain: It is composed of chips and other fragments in which the original form of plant tissue structure is preserved, commonly has fibrous texture with a very dull luster. Fusain is very friable and resembles charcoal. Commonly, it is concentrated in bedding layers or lenses that form planes of weakness in coal and thus is often exposed on bedding surfaces of broken coal. The many pores (cell cavities and cracks) of fusain are sometimes filled with mineral matter.

The breakage and strength characteristics of lithotypes as defined by Falcon and Falcon (1987) are given in Table 4.1.

It is also worth noting that a detailed program of research work has been carried out at the Chamber of Mines, South Africa, on the relationship of coal petrology to coal cuttability. MacGregor (1983), in one of his published

TABLE 4.1

Breakage and Strength Characteristics of Lithotypes

Item	Vitrain	Clarain	Durain	Fusain
Description	Bright, shining vitreous	Bright to semi-bright laminated bands	Dull, dark grey to black, massive compact to granular, lusterless structures	Flat, fibrous, bright rectangular lenses
Breakage characteristics	Brittle, cubic blocks, angular fragments	Semi-brittle to shattering and shearing	Inelastic, dense, solid, hard, tough, resistant to breakage	Friable, soft, easily pulverized to dust
Major microlithotypes	Vitrite Clarite	Clarite Vitrinertite Trimacerite Durite	Durite Inertite Inertoderite Carbominerite	Fusite Semifusite
Major macerals	Vitrinite	Mixed	Inertinite (Not Fusinite)	Inertinite (Only Fusinite)
Crushability de-ashed ratio	3.8	5.1	13.6	1.8
Density (g/cm³)	1.3–1.4	1.3–1.5	1.5–1.7	1.0–1.25
Strength categories of jeremic	Soft Vitrain	Semi-hard Clarain and composite of vitrain, clarain, and durain	Hard Durain shaley coal, coaly shale	Very soft Fusain
Cleavage	Intensive regular	Irregular erratic	Traces or none, uncleated	Open fractures
Uniaxial compressive strength (kg/cm²)	10–70	70–180	200–400	0–10
Borehole-core breakage	Often fragmented	Discs and half cylinders	Cylindrical core, height greater than diameter	Fines
Breakage on uniaxial compression	Cracking, compaction, yielding	Elasto-plastic failure	Elasto-brittle failure	Flow deformation and pulverization
Sizing after compression	Fine to nut size	Small lumps	Large lumps	Very fine sizes

Source: Falcon, L.M., and Falcon, R.M.S., *J. South Afr. Inst. Min. Metall.*, 87, 323–336, 1987.

works, concludes that certain variables, such as percentage volatiles and vitrinite, show some degree of correlation with certain mechanical properties (Hardgrove Grinding index) and the cuttability of South African coals. The study also endorses the view by Mackowsky (1967) that the breaking and grinding properties of coal do not depend on the hardness, which are characteristics of a homogenous substance, but on the strength, which characterizes the mechanical behavior of heterogeneous substances (MacGregor 1983, MacGregor and Barker 1985). Breakage and strength characteristics of lithotypes are summarized by Falcon and Falcon (1987), as given in Table 4.1. Some of the above section has already been published by Bilgin and Phillips (1994).

4.3 Coal Characterization through Vicker's Microhardness and Shore Scleroscope Measurements

As explained in the previous section, a coal seam is formed by a series of macroscopic bands known as lithotypes, for example, vitrain, durain, clarain, and fusain. Each lithotype has a particular mechanical strength, which causes structural heterogeneity in banded coal seams, and these are related directly to the strength of the seam (Falcon and Falcon 1987). However, it is almost impossible to measure the strength of each of these thin small bands with conventional testing methods. Vicker's microhardness and Shore Scleroscope hardness offer a good opportunity to measure the hardness or the strength of each band which also help to identify each lithotype.

4.3.1 Vicker's Microhardness

Vicker's microhardness is determined by pressing a diamond indenter with a certain static load into the maceral under a microscope. The greater the indentation depth is, the lower the microhardness of the coal. The value of the microhardness is expressed by the load applied per contact area between the indenter and the coal (in kilograms per square millimeter). Although the microhardness is the standard testing parameter of metals, ceramics, and composites, it is also used on coal specimen especially for identification of lithotypes as it will be explained below.

Works done by Nandi et al. (1977) demonstrated the practical use of Vicker's microhardness in coal petrology. Different types of microhardness values and indents were obtained on different ranks of coal showing that oxidation transformed the plastic state of fresh vitrinite into an elastic state. The transformation occurred rapidly in high volatile coals, but more severe oxidation was required to cause this change in low volatile coals. Reflectance also increased with oxidation. Reflectance measures the amount of light that is reflected from a polished piece of vitrinite, and it is in the range 0.6%–1.8%

for coking coals (range of bituminous coals). The coals with the lowest reflectance have the lowest rank and highest volatile matter. The microhardness data combined with the reflectance data were capable of defining the boundary where rapid transition from the plastic to elastic state occurred. Movement of this boundary made it possible to follow the oxygen penetration into coal particles.

Macmillian and Rikerby (1979) in their extensive studies noticed that none of the coals studied showed any significant difference between the hardness of a surface oriented perpendicular to the bedding plane and one parallel to it, regardless of the choice of load and loading time used in the measurement, and regardless of whether the indentations are placed in the dominant vitrite microlithotype or at random. Hence, despite the microstructural heterogeneity and anisotropy of the typical coal, its hardness can adequately be defined by a single number. It was also apparent that the hardness of coal was influenced much more by the extent of coalification than by the nature of its precursors since the scatter in the hardness data obtained by indenting at random is small compared to the variation of hardness with rank.

Das (1985) suggested that an index of brittleness could be established from the morphological analysis of the Vicker's microhardness impressions, as seen in Figure 4.1. Table 4.2 is a summary of Das's results for coal characterization based on Vicker's microhardness tests.

Mukherjee et al. (1989) used the Vicker's microhardness test on 50 Gondwana coal samples from India, representing the entire coal series from peat to

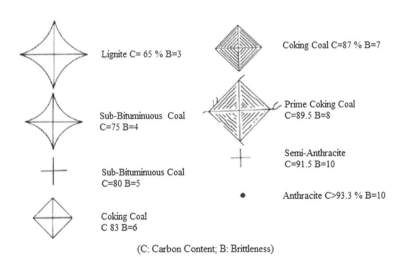

Lignite C= 65 % B=3

Sub-Bituminuous Coal C=75 B=4

Sub-Bituminuous Coal C=80 B=5

Coking Coal C 83 B=6

Coking Coal C=87 % B=7

Prime Coking Coal C=89.5 B=8

Semi-Anthracite C=91.5 B=10

Anthracite C>93.3 % B=10

(C: Carbon Content; B: Brittleness)

FIGURE 4.1
Indents of diamond indenter in Vicker's microhardness tests when testing different coals. (From Das, B., The microhardness technique and its application to coal and coal mining, Division Report, ERP/CRL 85-51 (OP), Canada Center for Mineral and Energy, Technology, Federal Department of Energy Mines and Resources, Calgary, Canada, 1985.)

TABLE 4.2

Geotechnical Classification of Coal Based on Vicker's Tests

Group	Geotechnical Properties	Typical Coal	Geotechnical Characteristic
I	HV < 20 B < 4	Lignites, low rank bituminous coals	Soft and ductile coal, difficult to cut, no cracks, less fines or dust, low abutment pressure
II	HV = 40 B = 5	Low volatile sub-bituminous and high volatile bituminous coals	Hard and elastic coals, blocky pieces, less cracks, less fines, cut burst prone
III	HV < 25 B = 7–8	Coaking coals	Soft friable coals, highly fractured, high amount of fines, higher permeability, and methane emission rate
IV	HV > 60 B = 9–10	Semi-anthracites and Anthracites	Very hard and brittle coals, dusty, but not to an extreme degree, high abutment pressure, highly prone to rock burst

Source: Das, B., The microhardness technique and its application to coal and coal mining, Division Report, ERP/CRL 85-51 (OP), Canada Center for Mineral and Energy, Technology, Federal Department of Energy Mines and Resources, Calgary, Canada, 1985.

Note: HV = Vicker's microhardness (kg/mm^2); B = Brittleness index.

semi-anthracite. The tests were carried out on the huminite/vitrinite maceral, supplemented by reflectance measurements. Observations involving over 5000 measurements have established the relation between microhardness and the rank of coal. A maximum value of microhardness is found at 0.80% reflectance, corresponding to 82.0% carbon content and a minimum at 1.68% reflectance, corresponding to 91.0% carbon content. The authors concluded that this type of measurement might form the basis for determination of the true hardness characteristics of a coal and serve as a useful physical parameter for the selection of coal for specific industrial use. On the other hand, the Vicker's microindentation hardness is a very sensitive measurement, which can be carried out also selectively on any of the lithotypes of coal. The trend of microhardness variation in relation to rank has been established for the vitrinite/huminite maceral, further measurements on the other components, followed by computation with respect to the volume percentage of the overall coal, is likely to reveal the true hardness characteristics of any coal.

Investigation carried out by Kozusnikova (2009) on the determination of the microhardness and elastic modulus of coal lithotypes by using Vicker's indentation test fortifies the views that this method is well suited for identifying coal lithotypes giving different indentation character, different hardness, and elasticity modulus values as given in Table 4.3 and Figure 4.2.

Kossovich et al. (2016) reported the test data on continuous indentation applied to different rank coal and anthracite on the surfaces oriented in

TABLE 4.3

Elasticity Modulus, Indentation Values and Character of Indentation for Different Lithotypes

Maceral Group	Elasticity Modulus (GPa)	Vicker's Hardness, HV_{50} (kg/mm^2)	Character of Indent
Vitrinite	6.20	82.68	Pyramid
Inertinite	7.11	94.4	Cross
Liptinite	4.43	55.66	Cross

Source: Kozusnikova, A., *GeoLines,* 22, 40–43, 2009.

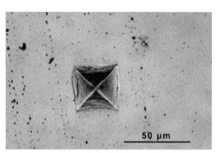

Indent on vitrinite F=0.5N Ro=0.87 %   Indent on liptinite F=0.5N

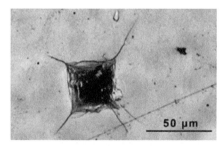   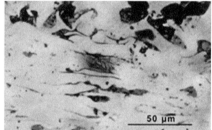

Indent on vitrinite F=0.5N Ro=1.23 %   Indent on inertinite F=0.5N

(F: Force; Ro: Reflectance)

FIGURE 4.2
Indents of diamond indenter in Vicker's microhardness tests when testing different lithotypes. (From Kozusnikova, A., *GeoLines,* 22, 40–43, 2009.)

parallel and perpendicular to bedding. Differences of mechanical properties and images of indents were illustrated for lithotypes represented by vitrinite and inertinite. Under the tests on the bedding-parallel surface, the lithotypes of coal and anthracite behaved more elastically than on applied perpendicular to bedding plane. This investigation showed also localization of anomalous stress concentration that might end with the initiation and growth of fractures and allowed estimation of voids, cavities, etc., which were the potential collecting volumes of methane. In the long view, this approach showed the applicability of this method for prediction of gas content of coal seams.

4.3.2 Shore Scleroscope Rebound Hardness

Klawitter et al. (2015) tested samples from coal seams of the Bowen Basin in Queensland, Australia. Samples were selected to include different coal ranks and lithotypes. The tests were carried out unconfined on slabbed coal core and confined on epoxy encased coal blocks used for coal petrographic study. They concluded that the unconfined samples showed that the hardness varied with lithotype. It increased with decreasing amount of bright bands or with coal rank having thermal maturity. The test results of the confined samples showed little variation with lithotype, but showed a parabolic correlation of hardness with rank, similar to the behavior found with Hardgrove Grindability tests. The resulting fractures of the Shore Scleroscope rebound hardness test were analyzed under the microscope to understand the fracture pattern, which could be scaled up to understand fracture propagation in natural systems and when induced in gas reservoirs. They also concluded that this test could be used without too much effort and expenses during the exploration to assist and estimate relationships to rank and type and to obtain empirical correlations between other coal parameters (Klawitter et al. 2015).

4.4 Conclusive Remarks

Coal is a non-homogeneous material which is composed of a number of microscopic organic and inorganic constituents. Macerals and minerals occur together in various proportions, forming macroscopic and microscopic bands, which are called microlithotypes. This chapter clarifies the characteristics of these lithotypes in relation to contribution to fracture and cuttability mechanism of coal.

Each lithotype has a particular mechanical strength, which causes structural heterogeneity in banded coal seams, and these are related directly to the strength of the seam (Falcon and Falcon 1987). The frequency of fissures is highest in vitrinites and the layers of vitrain. This is the reason why in run-of-mine coal, durites and most of the trimacerites are concentrated in the coarser size fraction (Stach et al. 1982). During mining, it has been shown that the energy required to break vitrain, clarain, and durain was two, three, and seven times higher, respectively, than that of the required to break fusain. It has also been found that the amount of power required to mine tough durain coal by a continuous miner was as much as 40% higher than that of the required for a friable clarain coal (Falcon 1978a, 1978b).

Vicker's microhardness and Shore Scleroscope hardness values are found to differentiate the macerals group for strength and cuttability characterization of the coal. Das (1985) suggested that an index of brittleness could be established from the morphological analysis of the Vicker's microhardness impressions.

At higher rank, the microhardness increases. The authors concluded that this type of measurement might form the basis for determination of the true hardness characteristics of coal and serve as a useful physical parameter for the selection of coal for specific industrial use. On the other hand, the Vicker's microindentation hardness is a very sensitive measurement, which can be carried out also selectively on any of the microcomponents of coal. The trend of microhardness variation in relation to rank has been established for the vitrinite/huminite maceral (Mukherjee et al. 1989).

Investigation carried out by Kozusnikova (2009) on the determination of the microhardness and elastic modulus of coal lithotypes by using Vicker's indentation test fortifies the views that this method is well suited for identifying coal lithotypes giving different indentation character, different hardness, and elasticity modulus.

It is also concluded that Vicker's indentation test can be used without too much effort and expenses during the exploration to assist and estimate relationships to rank and type and to obtain empirical correlations between other coal parameters (Klawitter et al. 2015).

References

ASTM D 121-09. Standard Terminology of Coal and Coke.

Bilgin, N. and Phillips, H. R. 1994. Mechanical properties of coal. In: *Coal, Resources, Properties Utilization, Pollution*. Editor: Orhan Kural, Istanbul.

Das, B. 1985. The microhardness technique and its application to coal and coal mining. Division Report, ERP/CRL 85-51 (OP), Canada Center for Mineral and Energy, Technology, Federal Department of Energy Mines and Resources, Calgary, Canada.

Falcon, L. M. and Falcon, R. M. S. 1987. The petrographic composition of Southern African coals in relation to friability, hardness and abrasive indices. *Journal of South African Institute of Mining and Metallurgy*, 87(10):323–336.

Falcon, R. M. S. 1978a. Coal in South Africa, Part II: The application of coal petrography to the characterization of coal. *Minerals Science Engineering*, 10(1):28–52.

Falcon, R. M. S. 1978b. Coal in South Africa, Part III: The fundamental approach to the characterization and rationalization of South Africa coal. *Minerals Science Engineering*, 10(2):130–153.

Klawitter, M., Esterle, J., and Collins, S. 2015. A study of hardness and fracture propagation in coal. *International Journal of Rock Mechanics and Mining Sciences,* 76:237–242.

Kossovich, E. L., Dobryakova, N. N., Epshtein, S. A., and Belov, D. S. 2016. Mechanical properties of coal microcomponents under continuous indentation. *Journal of Mining Science,* 52(5):906–912.

Kozusnikova, A. 2009. Determination of microhardness and elastic modulus of coal components by using indentation method. *GeoLines,* 22:40–43.

MacGregor, I. M. 1983. Preliminary results on the relationship of coal petrology to coal cuttability in some South African coal. *Special Publication of Geological Society of South Africa,* 7:117–128.

MacGregor, I. M. and Baker, D. R. 1985. A preliminary investigation of prediction of the cutting forces for some South African coals. *Journal of South African Institute of Mining and Metallurgy,* 85:259–272.

Mackowsky, M. T. 1967. Progress on coal petrology. *Proceeding of the Symposium on the Science and the Technology of Coal.* Ottawa, Canada, pp. 60–78.

Macmillian, N. H. and Rikerby, D. G. 1979. On the measurement of hardness in coal. *Journal of Materials Sciences,* 14:242–246.

Mukherjee, A. K., Alam, M. M., and Ghose, S. 1989. Micro-hardness characteristics of Indian coal and lignite. *Fuel,* 68(5):670–673.

Nandi, B. N., Ciavagli, L. A., and Montgomery, D. S. 1977. The variation of the micro-hardness and reflectance of coal under conditions of oxidation simulating weathering. *Journal of Microscopy,* 109:93–103.

Stach, E., Mackowsky, M-T., Teichmüller, M., Taylor, G. H., Chandra, D., and Teichmüller, R. 1982. *Stach's Textbook of Coal Petrology.* 3rd revised and enlarged edition. Gebruder Borntraeger, Berlin, Germany, 535 p.

5

Physical and Mechanical Characteristics of Coal

5.1 Introduction

The mechanical properties of coal are of great importance in coal mining engineering where they relate to the selection of mining methods, determining the blasting parameters, stability analysis of coal roadways, prevention of coal bumps and bursts, dust generation, cuttability and workability or grindability of coal, slope stability analysis, etc. Due to this fact, the research carried out worldwide on the subject will be summarized.

5.2 Coal Density

Density of coal is true density as measured by helium displacement or is the mass divided by the volume occupied by the actual pore free solid in coal. The density of coal shows a notable variation with the rank of carbon content. Anthracite has a density about 1.55 g/cm³, bituminous coal about 1.35 g/cm³, and lignite about 1.25 g/cm³.

Speight (2015) claims that the standard test method ASTM D167 (2012) for determining the true density of coke may be applied also to coal.

5.3 Porosity and Permeability

Coal bed methane (CBM) is one of the unconventional energy sources, where many countries have started to explore seriously. In recent years, CBM projects have rapidly increased. Australia had no CBM production in 1995, but in 2008 extracted 4 billion m³ from its extensive underground coal reserves. China had in excess of 1.4 million m³ of CBM production in 2006. These amounts are small compared to United States (US) production of 61 billion m³ in 2007, more than 10% of the US domestic natural gas supply. However, all this production is significant since it comes from an energy resource that was barely utilized before 1985 (Al-Jubori et al. 2009).

Zhang et al. (2015) stated that the utilization of CBM could not only add to the world's gas supply, but also helps to enhance underground safety for mining workers and contribute to greenhouse gas. In wet CBM reservoirs, the cleats are usually initially saturated with water. During the primary recovery of CBM, water must be drained prior to gas production so that the reservoir pressure can be lowered and subsequent gas desorption from internal matrix surfaces can be initiated. Once desorbed gas enters the cleats (through diffusion) and achieves irreducible gas saturation, simultaneous gas and water flow occur in the cleats.

One of the main physical parameters governing gas emission from coal is permeability, which is defined as the resistance of strata to the passage of gas through it. Highly permeable coal offers good opportunity to recover methane. The unit of measuring the permeability is the Darcy, normally expressed in milli-Darcies (mD) which represents the flow capacity required for 1 mL of fluid to flow through 1 cm² for a distance of 1 cm when 1 atmosphere of pressure is applied.

Determination of coal cleat porosity, permeability, and gas-water relative permeability is necessary for prediction of methane production rates in coalbed metha.ne operations. Carbon sequestration in coal gas reservoirs, a technique to combat atmospheric CO_2, while simultaneously enhancing the recovery of methane (CH_4), is a viable option in the immediate future. However, injected CO_2 in deep coal seams, while being adsorbed, results in swelling of the solid coal matrix in addition to displacing additional methane. Among the various geological carbon sequestration options, permanent storage of CO_2 in coal seams is attractive due to their ability to enhance the recovery of coalbed methane (Gosh 2011). The injection of CO_2 not only enhances CBM production, but also helps reducing the release of this greenhouse gas. For this reason, it is essential to understand well the porosity and the permeability of coal.

The absolute permeability of coal reservoirs changes significantly during gas production, often initially decreasing, but then increasing as the reservoir pressure and gas content are drawn down. It has also been observed to decrease markedly during CO_2 injection to enhance coalbed methane recovery. With increased experience with CBM production, it has become abundantly clear that the permeability of coal varies with continued production. The most dramatic examples of these are several producing reservoirs in the San Juan Basin, with a permeability increase of as much as 100 times (Mitra et al. 2012). During drawdown of a reservoir by primary production, effective stress is believed to increase, resulting in permeability reduction due to the closure of cleats. However, methane is stored in coal as sorbed gas and production leads to desorption of gas. This is accompanied by "matrix shrinkage," which is believed to open up the cleats, thus leading to increased permeability.

Gash et al. (1992) investigated in the laboratory the effect of cleat orientation and confining pressure on cleat porosity, permeability, and gas-water relative permeability on coal samples from Fruitland, New Mexico, they proved that permeability was the largest parallel to the bedding planes in the face cleat direction. At 6.9 MPa confining pressure, permeability parallel to the bedding planes in the face cleat direction was 0.6 mD–1.7 mD and in the butt cleat direction, 0.3 mD–1.0 mD. These compared with 0.007 mD measured perpendicular to the bedding planes. Confining pressure was found to have a significant effect on cleat porosity, permeability, and relative permeability.

Durucan et al. (2013) tested seven different anthracite and bituminous coal specimens from France, Germany, Scotland, and the United Kingdom and found that absolute permeability changed between 0.52 mD and 9.51 mD and porosity changed between 0.12% and 1.8% (±0.6%).

Ramandi et al. (2016) expressed that coal had a dual porosity system containing micropores and a network of natural fractures known as cleats. Cleat properties are affected by many factors including coal rank, layer thickness, composition, and in-situ stress regimes. Cleat network structure varies noticeably with coal composition and is among the most important factors affecting permeability in coal.

5.4 Uniaxial and Triaxial Compressive Strength

The compressive strength of coal is a most difficult property to investigate, since it may vary laterally as well as vertically in the same seam and may have one or more sets of discontinuities or planes of weaknesses, namely, face cleavage, butt cleavage, and bedding planes. Uniaxial compressive strengths of coals from different countries gathered from the published data are given in Table 5.1.

Mathey (2015) evaluated a number of 403 tests on the uniaxial compressive strength for the overall coal strength of South African seams and stated that the laboratory strength of South African coal seams was seen to cluster very closely around the average value of 22.3 MPa. Around 80% of all uniaxial compressive strength tests specimens were prepared from 60 mm diameter borehole cores, the rest came from 25 mm to 100 mm diameter cores drilled out of sample blocks. He did also a comparison in an international level and reported that the strength of coal from seams in India and

TABLE 5.1

Uniaxial Compressive Strength of Coals from Different Countries

Country	Cube Specimen Size (cm × cm)	Perpendicular to Bedding Plane (MPa)	Parallel to Bedding Plane (MPa)	References
England Pentremawr	2.56	38.3	34.4	Evans and Pomeroy (1966)
England Deep Duffryn	2.56	18.3	16.1	Evans and Pomeroy (1966)
England Oakdale	2.56	7.9	6.1	Evans and Pomeroy (1966)
England Preasly	2.56	57.4	39.2	Evans and Pomeroy (1966)
India, Hazaribagh	Cylinder 5 × 10	16.7	—	Anand and Gri (2015)
India Jharia	Cylinder	2.8–4.7	—	Kumar et al. (2015)
Brazil A.Sangao Mine	5.5 10.0 15.1 20.1 25.1 31.3	20.7 16.5 14.0 12.7 10.6 9.5	—	Gonzatti et al. (2014)
South Africa Witbank Coalfield	Cylinder 4.5 × 9	18.2	—	Roxborough et al. (1981)
USA Marker	5 25	48.3 20.5	—	Holland (1964)
USA Pittsburgh	5 25	27.6 13.8	—	Holland (1964)
USA Clintwood	5 25	24.1 10.3	—	Holland (1964)
Turkey, ELI, Darkale, +285/102	5	19.8 +/− 4.7	—	Bilgin et al. (1992)
Turkey, ELI, Darkale, +285–II	5	25.8 +/− 9	—	Bilgin et al. (1992)
Turkey, ELI, Eynez, +457	5	34.3 +/− 13.3	—	Bilgin et al. (1992)
Turkey, Milten, İstanbul	5	38.6 +/− 4.5	—	Yazici et al. (1998)
Turkey, Soma, Imbat Mine	5	22.3 +/− 14	—	Bilgin et al. (2011)
Turkey, Soma, Imbat, Upper Level	5	95.7 +/− 6.6	—	Bilgin et al. (2011)
Turkey, Amasra	5	46.2	—	Bilgin et al. (2010)

the US, however, showed a relatively large scatter of between 12.5 MPa and 42.5 MPa in India and between 10 MPa and 47.5 MPa in the US. The average laboratory tested coal seam strength was again very close to the South African value, for example, 23.5 MPa in the case of India and 22.1 MPa for the analyzed US coal seams.

It is important to note that the following factors may effect directly the compressive strength of coal and they should be taken into consideration for any engineering design.

5.4.1 The Rank or Petrology of Coal

According to Pan et al. (2013), the uniaxial compressive strengths of coals increase with increasing coal ranks because pore volumes, in general, decrease with increasing vitrinite reflectance. As vitrinite reflectance or coal rank increases, coal has less microporous structure and higher uniaxial compressive strength. He also pointed out that after different research results, a (U) shape relationship between carbon content and uniaxial compressive strength was found. A minimum value of 7 MPa was obtained for carbon contents of around 70%–80% and maximum uniaxial compressive strength values between 33 MPa and 55 MPa for carbon content around 60%.

5.4.2 Plane of Weaknesses, Bedding Planes, and Cleats

There is not much data available in the literature on the effect of bedding planes and cleats on compressive strength of coal. However, as it is seen in Table 5.2 according to Evans and Pomeroy (1966), there is a significant difference between uniaxial compressive strength of coal loaded perpendicular and parallel to bedding plane, the ratio being around 1.3.

TABLE 5.2

Effect of Bedding Plane on Indirect (Brazilian) Tensile Strength of British Coals

Country	Perpendicular to Bedding Plane (MPa)	Parallel to Bedding Plane (MPa)
Pentremawr, Pumpqart	1.58	2.41
Deep Duffryn, Gellideg	0.69	0.96
Oakdale, Meadow	0.48	0.76
Cwmtillery, Garw	0.55	0.96
Rossigton, Barnsley Hards	1.1	4.07
Teversal, Dunsil	1.1	3.0
Markham, Blackshale	0.62	1.72
Rossington, Barnsley Brights	0.76	2.48
Linby, High Main	0.9	2.9
Mean	0.86 ± 0.33	2.14 ± 1.06

Source: Evans, I. and Pomeroy, C.D., *The Strength, Fracture and Workability of Coal*, Pergamon Press Ltd., London, UK, 1966.

5.4.3 Dimensions of Cubes, Rectangular Samples, and Cores

Studies on coal mine pillar stability increased after unexpected major failures in the Coalbrook Colliery disaster, which occurred in South Africa in 1960. In a 20 minutes' period, over 4400 pillars collapsed, resulting in the death of 437 miners (Quinteiro et al. 1995). Therefore, the effect of specimen size of coal is of prime importance in designing pillars in coal mining, and there is immense published data on the subject. The reader is recommended reading the book published by Arioglu and Tokgoz (2011) on "*Hard Rock Mass and Coal Strength*" giving an intensive survey on pillar stability. Bearing in mind that this subject is beyond the scope of this chapter, only a brief summary on the size effect of test specimen on coal strengths will be given below.

Gaddy (1956) investigated the strength of cubical coal specimens as related to their size. He found that for coals having different characteristics, the relationship between the compressive strength of a cub specimen (σ_c) and its absolute size (D) could be expressed by Equation 5.1.

$$\sigma_c = K/D^{0.5} \tag{5.1}$$

where K is Gaddy constant depending on coal properties.

Bieniawski (1984) is one of the researchers who investigated the effect of the specimen size on the strength of coal. He used specimen of cube size up to 2 m, and he noticed that the effect of the size of the specimen was insignificant after 40 cm of cube size. His results are illustrated in Figure 5.1.

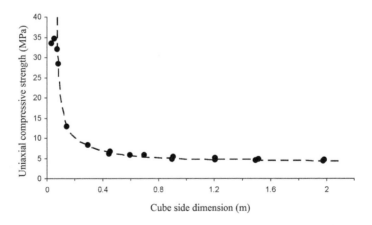

FIGURE 5.1
Variation of uniaxial compressive strength with cube size dimensions. (Adapted from Bieniawski, Z., *International Journal of Rock Mechanics and Mining Science* 5, 325–335, 1984 and published in Bilgin, N. et al., *Mechanical Excavation in Mining and Civil Industries*, CRC Press/Taylor & Francis Group, Boca Raton, FL, 2014a.)

However, it is obvious that the preparation of this size of specimen is practically very difficult for laboratory routine tests. This fact leads the investigators to use the cubes size of 4 cm–5 cm or Core of 54.7 mm in diameter (NX) size core specimens.

The correlation between the test results obtained from core and cube specimens is of prime importance for test standardization. This questionable fact was investigated by Townsend et al. (1977). They established a relationship between strengths of cubical and cylindrical specimens ranging in size from 2 square inches to 16 square inches and they found that the cubes were approximately 30% stronger than the cylinders over the size range tested, bearing in mind that the cubical concrete specimens are approximately 5%–20% stronger than the cylindrical specimens over the size range tested.

Udd et al. (1980) found that the strength of coal was related to the length to diameter ratio as given in Equation 5.2, where compressive strength (σ_c) is in MPa, length (L) in (cm), and diameter (D) in (cm).

$$\sigma_c = 8.934\,(L/D)^{-0.92} \tag{5.2}$$

The first pillar strength formula for South African coal was published by Salamon and Munro (1967), used a database of 27 collapsed and 98 stable pillar geometries, and derived a statistical maximum likelihood pillar strength equation under the argument that collapsed pillars must have a predicted safety factor close to one and stable pillars should have safety factors larger than one. This resulted in Equation 5.3.

$$\text{Strength} = \frac{7.2 \cdot w^{0.46}}{h^{0.66}} \tag{5.3}$$

where strength is in MPa, w is the pillar square width (m), and h is the pillar height (m). The formula has remained the most frequently applied pillar strength equation in South Africa. However, there are several new attempts worldwide to develop a new formula. They will not be discussed here, since it is beyond the scope of this chapter.

5.4.4 Confining Stress

Sometimes in a mine, coal is left in situ either temporarily or permanently in the form of regular pillars or barrier pillars in longwall mining. The designer wants to maximize the extraction ratio while maintaining adequate pillar sizes to assure safety and continued operation of the mine. The strength of the coal in the pillars is therefore an important factor in mine design. In such cases, triaxial post failure behavior of the pillars is of upmost importance.

A comprehensive triaxial testing program involving coal from West Virginia, Pennsylvania, New Mexico, Utah, Wyoming, and Montana was carried out by Atkinson and Ko (1977). Five of the coals were tested in a NX-sized cell, while the sixth was tested in a Core of 21.5 mm in diameter (EX)-sized cell and post-peak or residual data were obtained. A straight line fit yielding values of the Mohr constants (c) cohesion and (φ) internal friction angle were found to adequately fit both the peak and residual strength data. A constant relation between peak and residual angles of friction was observed. Atkinson and KO (1977) concluded that data presented indicated that the US coals had a higher angle of friction and a lower cohesion on the average than that of reported previously for British coals. This result would indicate that caution is advised in adopting British mine design practice and formulas to the US conditions. A linear fit yielding the Mohr strength parameters c and φ was adequate to express the strength behavior of the coal. The ratio of the residual angle of friction to the peak angle of friction for coal is approximately 0.8 for the coals tested. He reported for the US coals that the peak cohesion values changed between 2.5 MPa and 8.6 MPa with a mean value of 4.5 MPa, the residual cohesion values changed between 1.9 MPa and 4.3 MPa with a mean value of 2.9 MPa, the peak internal friction angle changed between 32.2° and 53.5° with a mean value of 45.1°, and the residual internal friction angle changed between 28.4° and 42.8° with a mean value of 37.0°.

Mathey (2015) reported the results of a total 198 coal specimens tested in triaxial compression at confinement levels between 1.5 MPa and 20 MPa. The tests covered eight different coal seams of four coalfields. Two thirds of all tests were performed at 2.5 MPa, 5 MPa, and 10 MPa confinement levels and provided mean Mohr-Coulomb cohesion and angle of internal friction values of 11.3 MPa ± 1.9 MPa and 34.9° ± 4.6°.

5.5 Point Load Strength

The point load (PL) testing apparatus is widely used in laboratory and field on cores or irregular pieces of rocks for estimation of the uniaxial compressive strength (Broch and Franklin 1972, Brook 1977, 1985). The specimen is loaded to failure between two standard conical platens, as seen in Figure 5.2. The uncorrected PL strength (I_s) is calculated as the ratio of failure load and equivalent core diameter (De). I_s must be corrected to the standard equivalent diameter (De) of 50 mm and called $I_{s(50)}$.

An attempt has been made to use the PL test on coal specimens in the Aegean Lignite Mine (Bilgin 1992). Cube specimens of 5 cm in size were loaded perpendicular to the bedding planes with conical platens of standard geometry. Special care was taken in performing and calculating $I_{s(50)}$ values as described by Brook (1977, 1985). A satisfactory correlation was found between uniaxial compressive strength of coal (5 cm cubes) and PL strength values, as seen in Figure 5.3 (Bilgin et al. 2014b).

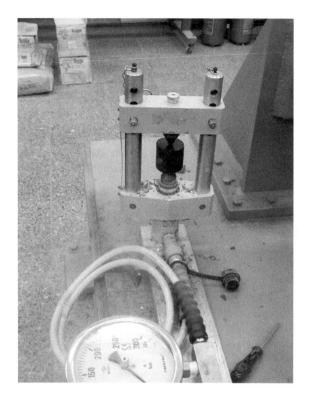

FIGURE 5.2
Point load test apparatus.

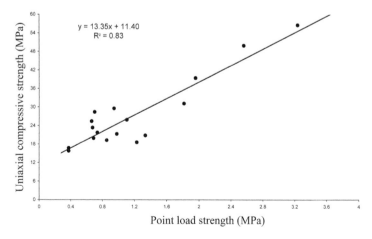

FIGURE 5.3
Relationship between point load strength and uniaxial compressive strength of coal specimens from Aegean Lignite Mine. (From Bilgin, N. et al., *Mechanical Excavation in Mining and Civil Industries*, CRC Press/Taylor & Francis Group, Boca Raton, FL, 2014a.)

5.6 Tensile Strength

It is essential to understand the factors effecting the tensile strength of coal since, as it will be explained within this book, tensile breakage plays an important role in coal cutting mechanics.

Evans and Pomeroy (1966) stressed that even when compressive stresses were applied to coal specimens, the breakage appeared to be essentially tensile in nature. The evidence of this was demonstrated by means of photoelasticity and applicability of weakest-link theory (Evans and Pomeroy 1966).

The obvious tests for tensile strength are the straight pull test and bending tests, however, the complexity of preparing samples for these tests necessitates of using Brazilian (indirect) test method for coal. The values found in the literature are mainly obtained from disc loading tests. Although the tensile stress is associated with an orthogonal compressive stress in disc testing, it is always preferred by researchers for its simplicity.

Brazilian tensile strength (σ_t) provides a measure of coal toughness, as well as strength. This parameter is usually measured using NX-sized core samples (54 mm in diameter) cut in length-to-diameter ratio of 0.5 and following the procedures of ASTM D3967 (2016) or International Society for Rock Mechanics (ISRM) (2007) suggested methods. The diameter is required to change less than 0.5 mm over the length of the sample and the core ends must be perpendicular to the core axis, up to a precision depending on the standard applied. The bedding plane and failure type are also the most important criteria for Brazilian tensile strength testing and should be noted before and after the testing on the report.

The Brazilian tensile strength is calculated by using Equation 5.4.

$$\sigma_t = 2 \cdot P/(\pi \cdot D \cdot L) \tag{5.4}$$

where:
  σ_t is the Brazilian tensile strength, MPa
  D is the diameter of the sample before testing, mm
  P is the maximum force on the sample before failure, N
  L is the length of the sample before testing, mm

The dynamic mechanical properties of coal are of great importance in coal mining engineering where they relate to the determination of cuttability and workability of coal, selection of blasting parameters, stability analysis of coal roadways under impact loading, and the prevention of coal bumps and bursts. The dynamic tensile strength is the subject of great interest, not only for the underground engineers (mining, geotechnical, and tunneling engineers), but also for other branches of engineering, such as mineral processing, where the main objective is associated with energy consumption during mineral/rock crushing and milling. From the mining engineering point of view, the interest with respect to the dynamic tensile strength of rock is very often associated with the fracture initiation, propagation, and failure that can develop in the pillars, roof, floor and caving zone, and in the gob during continuous mining. The underground rock and/or rock mass encounters dynamic or cyclic loading induced by drilling and blasting, mechanized extraction, and hydraulic support advance. Periodical dynamic rock mechanical properties should be considered and incorporated in the ground control and rock/coal burst control plans. To improve the mining safety conditions, dynamic rock properties need to be thoroughly investigated and incorporated in the mine design and ground control plans (Zhang and Zhao 2014).

Direct tensile tests are very difficult and complicated to perform. Thus, researchers developed several indirect tensile testing methods. Among them, the Brazilian disc specimen in the split Hopkinson pressure bar (SHPB) system is widely employed to achieve the dynamic tensile strength of coal or rock (Zhang and Zhao 2014). The indirect Brazilian disc methods provide a convenient means of conducting tests in terms of specimen manufacturing, experimental setup, and data reduction, but the complex networks of bedding planes in coals result in difficulties to characterize the dynamic features of coals. In the dynamic indirect tensile test, the coal specimen is placed between the incident bar and transmission bar. During the test, the striker bar is launched by the gas gun at high speed to impact the incident bar. The resulting strain wave $\varepsilon_{I(t)}$ is of short duration, reflected at the interface between the specimen and incidental bar, and recorded both as a reflected strain wave $\varepsilon_{R(t)}$ and as a transmitted strain wave $\varepsilon_{T(t)}$. The characteristics of these strain waves are obtained from two dynamic strain gauges mounted on each of the incident bar and transmission bar. Then, dynamic tensile strength of the specimen is calculated as explained by Zhao et al. (2014).

One using tensile strength of coal should be careful and take into account of the following factors used in obtaining test values.

1. Testing method, direct (direct pull), bending, indirect (Brazilian method or dynamic testing methods),
2. Effect of the bedding planes and presence of cleats,
3. The effect of specimen size, and
4. Dry or saturated.

Results of the Brazilian tensile strength tests given in Table 5.3 reflect the importance of bedding planes (Evans and Pomeroy 1966). Tensile strength parallel to bedding plane was reported to be 2.5 times higher than the values for perpendicular to bedding plane. The tests were obtained by using 2.54 cm diameter discs having thickness of 0.8 cm.

The effect of bedding planes is also reflected in the studies reported by Zhao et al. (2014, 2016) as it is seen in Table 5.3. In that study, a number of dynamic indirect tensile tests for Datong coal (China) were conducted

TABLE 5.3

Effect of Bedding Plane, Saturated and Dry Conditions on Dynamic
Tensile Strength of Coals from Datong Mine China

Bedding Plane (Degrees)	Dynamic Tensile Strength, Dry (MPa)	Dynamic Tensile Strength, Saturated (MPa)
0	4.92 ± 0.34	4.04 ± 0.66
22.5	4.61 ± 0.51	4.67 ± 0.32
45	5.29 ± 0.50	5.36 ± 0.27
67.5	5.10 ± 0.64	5.27 ± 0.36
90	4.62 ± 0.69	4.41 ± 0.52

Source: Zhao, Y. et al., *Int. J. Coal. Geol.,* 132, 81–93, 2014; Zhao, Y. et al.,
        Rock Mech. Rock. Engg., 49, 1709–1720, 2016.

by using an SHPB in dry and saturated samples. It focused on the influence of the bedding structure on the dynamic indirect tensile strength of Datong coal. Coal specimens with different bedding directions (along the loading direction) were tested by using the SHPB under different impact velocities. Before the test, the density and dynamic elasticity modulus were measured. To further define the influence of bedding structure, x-ray micro computed tomography is used to provide spatial data to inform the meshing of discrete element-based numerical models to investigate the influence of different bedding structures on the dynamic tensile failure of coal. According to the ISRM suggested method (Zhou et al. 2012), the diameter and height of these specimens were 50 mm and 25 mm, respectively. However, it is interesting to note here that the effect of bedding plane was not as strong as the values obtained for British coal, which might be due to difference lying in the bedding characteristics of two sets of coal specimens.

Tables 5.4 and 5.5 may serve as a guide to compare the tensile test results obtained from direct and indirect tests. As it is seen from these two tables, the indirect tensile strength values are 4.1 times higher than the direct pull test results.

TABLE 5.4

Direct Tensile Strength Test Results of a Chinese Anthracite

Boring Direction	Number of Samples	Uniaxial Tensile Strength (MPa)	Coefficient of Variation (%)
X	4	0.42	69
Y	5	1.04	71
Z	6	0.57	76

Source: Okubo, S. et al., *Int. J. Coal Geology,* 68, 196–204, 2006.

TABLE 5.5

Indirect Tensile Strength Test Results (Brazilian test) of a Chinese Anthracite

Boring Direction (Loading Direction)	Direction of Tensile Stress	Number of Samples	Indirect Tensile Strength (MPa)
Y(Z)	X	6	2.1
Z(Y)	X	10	1.9
X(Z)	Y	8	3.7
Z(X)	Y	5	2.8
Y(Z)	Z	8	3.2
Z(Y)	Z	5	3.0

Source: Okubo, S. et al., *Int. J. Coal Geology*, 68, 196–204, 2006.

Table 5.6 gives the comparison of tensile strength test results given by different researchers. Tests were carried out on NX core samples loaded perpendicular to bedding planes.

Mathey (2015) tested 186 disc specimens for indirect tensile strength of coal from four coalfields, ten collieries, and ten different seams from South Africa.

TABLE 5.6

Indirect Tensile Strength Test Results Carried Out on NX Cores Perpendicular to Bedding Planes, from Different Countries

Country	Indirect Tensile Strength MPa ± Standard Deviation	References
Australia, Bulli Seam	1.85 ± 0.86	Roxborough and Hagen (2010)
Australia, Young Wallsend Seam	1.26 ± 0.62	Roxborough and Hagen (2010)
Australia, Whybrow Seam	1.56 ± 0.65	Roxborough and Hagen (2010)
Australia, Great Northern Seam	3.63 ± 1.40	Roxborough and Hagen (2010)
India, Talcher Coalfields	1.5 ± 0.41	Anand and Gri (2015)
South Africa, 186 specimen	1.6 ± 0.72	Mathey (2015)
Turkey, Eynez	5.67	Copur and Balci (2010)
Turkey, Eynez	2.08	Copur and Balci (2010)
Turkey, Amasra	0.57 ± 0.2	Bilgin et al. (2010)
Turkey, Amasra	0.20 +/− 0.1	Bilgin et al. (2010)
USA, Beehive	2.10	Atkinson and Ko (1977)
USA, Bruceton	1.47	Atkinson and Ko (1977)
USA, Decker	5.1	Atkinson and Ko (1977)
USA, Federal No 2	2.16	Atkinson and Ko (1977)
USA, Hanna	0.97	Atkinson and Ko (1977)
USA, York Kanyon	0.27	Atkinson and Ko (1977)

The samples exhibited a lognormal frequency distribution, only tensile strength values below 0.5 MPa were not well represented by the lognormal distribution. The average indirect tensile strength of the samples was 1.60 MPa with a standard deviation of 0.72 MPa. He found also that the average ratio of compressive strength to tensile strength was around 15.

The ratio of compressive strength to tensile strength is of upmost interest for researchers and practicing engineers. Evans and Pomeroy (1966) in a comparative study found that the ratio of compressive strength perpendicular to bedding plane and tensile strength parallel to bedding plane was 17 and the ratio of compressive strength parallel to bedding plane to tensile strength perpendicular to bedding plane was 28. However, it is obvious at the rank of coal, the existence of bedding planes and the cleats, the specimen size will effect this ratio. A typical example to this is the results obtained by Anand and Gri (2015). They found a ratio of 11.1 between compressive strength and tensile strength of coal using core specimens of NX size.

5.7 Static and Dynamic Elasticity Modulus

Young's modulus of coal is a critical parameter in the design of coal mining and coalbed methane development, horizontal borehole stability, enhanced methane recovery, stress-dependent permeability in coalbed methane wells, coal pillar stability, coal outburst, and groundwater inrush in underground mining and so on (Pan et al. 2013).

Mathey (2015) tested a total number of 322 specimens from 5 different South African coalfields and 11 different seams. Around 80% of the test specimens were prepared from 60 mm diameter borehole cores, the rest came from 25 mm to 100 mm diameter cores drilled out of sample blocks. The values were determined as both the tangential and secant moduli at 50% of the peak stress. The average of tangential elasticity modulus of coal specimens was found to be to 4.45 GPa with a standard deviation of 2.04 GPa.

Evans and Pomeroy (1966) reported that for cube specimen size of 3.81 cm, Young's modulus of Barnsley hard coals (United Kingdom) was 2.7 GPa for perpendicular to bedding plane and 3.3 GPa for parallel to bedding plane, and Young's modulus of anthracite perpendicular to bedding plane was 3.8 GPa and parallel to bedding plane was 4.0 GPa. However, these values were found to be 20% higher for dynamic Young's modulus for the same specimen.

Szwilski (1984) stated that the modulus of elasticity in the loading direction being parallel to the bedding plane was larger (stiffer) than the modulus perpendicular to the bedding plane. According to him, there were two principal features to consider. Firstly, the degree of stiffness increased with confinement pressures, secondly, the elasticity constants changed as a function of the percentage of carbon or rank of coal. There was a maximum value at about 80%–85% carbon and a minimum at about 90% carbon.

Morcote et al. (2010) reported that laboratory ultrasonic velocity measurements of different types of coal and their dynamic elasticity properties depended on coal rank and applied effective pressure. Coal thermal maturity had a significant influence on dynamic elasticity properties of coal. Bituminous coal had lower velocities than that of semi-anthracite and anthracite, the latter had the highest rank and velocities. Dry bulk and dry shear moduli increased with increasing coal rank, whereas the ratio of V_P to V_S (acoustic P and S wave velocities) decreased with increasing coal rank. Coal acoustic wave velocities also depended on confining pressure. The dependence of velocities on confining pressure was greater at lower pressures up to 5 MPa, which might be due to the presence of micro cracks, above this pressure, changes in velocities were minimal.

Pan et al. (2013) emphasized that the compressive strength and Young's modulus of coal were closely related to its physical and chemical properties, which were, in turn, governed by coal type and rank. They increased with increasing coal ranks depending on pore volumes, in general, decreased with increasing vitrinite reflectance. As vitrinite reflectance or coal rank increased, coal had less microporous structure and higher uniaxial compressive strength and Young's modulus. Therefore, using vitrinite reflectance value instead of vitrinite content was advantageous for correlating to the rock strength. Based on data for different coal ranks (bituminous to anthracite) in three coal basins in China (Qianqiu, Tangshan, and Yanma collieries), the correlation given in Equation 5.5 was obtained.

$$\sigma_c = 3.3844 \cdot e^{0.4744 \cdot E} \tag{5.5}$$

where σ_c is uniaxial compressive strength in megapascal and E is elasticity modulus in megapascal.

Durucan et al. (2013) in their study on "Two Phase Relative Permeability of Gas and Water in Coal for Enhanced Coalbed Methane Recovery and CO_2 Storage" tested six different anthracite and bituminous coal specimens from France, Germany, and Scotland and found that Young' modulus changed between 1.8 GPa and 3.9 GPa with a mean value of 2.68 ± 0.7 GPa.

Bilgin et al. (2011) found a linear relationship between static Young's modulus and uniaxial compressive strength of 5 cm cube specimens for a Turkish lignite mine, as given in Figure 5.4, the specimens were loaded perpendicular to bedding plane. The relationship between static and dynamic Young's modulus for the same coal field is given in Figure 5.5. As seen, the static elasticity modulus is around three times higher than the dynamic elasticity modulus, which might be due to the inhomogeneity of the coal specimens and complex cleat systems existing within the coal specimen.

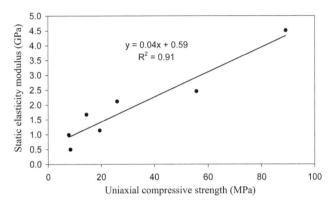

FIGURE 5.4
Relationship between elasticity modulus and uniaxial compressive strength of coals from Soma-Eynez Coal Field. (From Bilgin, N. et al., The cuttability and cavability of thick coal seam in Soma-Eynez coal field IR-75153 of TKI operated by IMBAT AS, Project report, Istanbul Technical University, 59, 2011.)

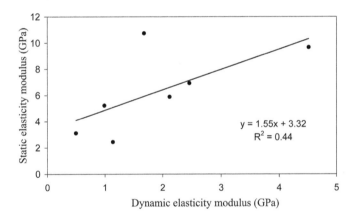

FIGURE 5.5
Relationship between static elasticity and dynamic elasticity modulus of coals from Soma-Eynez Coal Field. (From Bilgin, N. et al., The cuttability and cavability of thick coal seam in Soma-Eynez coal field IR-75153 of TKI operated by IMBAT AS, Project report, Istanbul Technical University, 59, 2011.)

5.8 Poisson's Ratio

The ratio of lateral strain to longitudinal strain is called Poisson's ratio or μ. Poisson's ratio of coal is used to predict the sonic velocities of coal underground. However, the measurement of elasticity constants of coal presents many difficulties such as preparation of samples, dependency on inhomogeneities, orientation of discontinuities, chemical properties, humidity, etc., Krey (1963), who pioneered in-seam seismic methods for coal, estimated the longitudinal velocity of coal to be VL = 1.2 km/sec and used a value of μ = 0.25.

Szabo (1981) made an extensive research on coal Poisson's ratio on 12 samples having density changing between 1.3 g/cm³ and 1.75 g/cm³ and reported mean values in XY direction of 0.26, in XZ direction of 0.35, and in YZ direction of 0.43. He also concluded that a Poisson's ratio of 0.346 is reasonably representative of a wide range of coal grades. In the absence of the availability of more detailed elasticity data, this value may be used for estimating acoustic velocities of coals. It is interesting to note that Mathey (2015) reported the average Poisson's Ratio for the 290 specimens of South African coal as 0.306 with a standard deviation of 0.147.

5.9 Impact Strength Index

The impact strength index (ISI) test is described in detail by Evans and Pomeroy (1966) and used for determining some basic parameters of coal. The apparatus consists of a vertical hollow cylinder with 4.45 cm internal diameter. A steel plunger with a mass of 1.8 kg fits loosely inside the hollow cylinder. The coal specimen of 100 g in the 9.5 mm–3.2 mm size range is poured into the cylinder and the plunger is dropped 20 times into the cylinder from a height of 30.5 cm. Finally, the coal specimen is removed from the apparatus and sieved. The mass of coal in grams retained on the 9.5 mm–3.2 mm sieve is defined as the impact strength index of the coal. The impact strength of the coal is related to compressive strength in an exponential function, as given in Figure 5.6 (Bilgin et al. 2014).

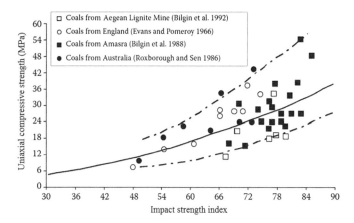

FIGURE 5.6

Relationship between impact strength index and uniaxial compressive strength of coals. (From Bilgin, N. et al., *Mechanical Excavation in Mining and Civil Industries*, CRC Press/Taylor & Francis Group, Boca Raton, FL, 2014a.)

TABLE 5.7

Classification of Strength of
Zonguldak Hard Coals According
to Impact Strength Index (ISI)

ISI Range	Strength Class
ISI > 75	Extremely hard
75 < ISI < 60	Very hard
60 < ISI < 40	Hard
ISI < 40	Soft

Source: Bilgin, N. et al. (1988).

The strength of Turkish hard coal seams from Zonguldak was classified based on impact strength index after Bilgin et al. (1988), as given in Table 5.7.

The relationship between volatile matter, uniaxial compressive strength, and impact strength index values as previously published by Speight (2015) is seen in Figure 5.7. Uniaxial compressive strength and impact strength index values decreased with volatile matter content showing a minimum value for volatile matter of 25% in hard coals and strength increased thereafter. Coal samples collected from Imbat Mine, Soma, showed also the same trend as seen in Figures 5.8 and 5.9 for volatile matter giving a minimum at around 40%, which showed clearly that the origin of coal effected the relationships between coal strength and coal chemical properties.

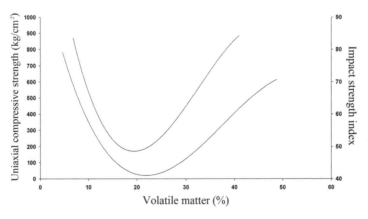

FIGURE 5.7
Relationship between volatile matter, uniaxial compressive strength, and impact strength index. (From Speight, J., *Handbook of Coal Analysis*, Wiley, New York, 2015.)

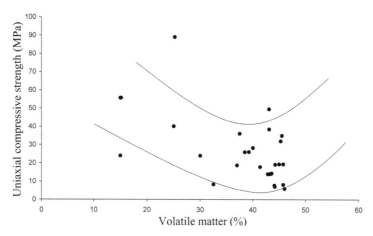

FIGURE 5.8
Relationship between volatile matter and uniaxial compressive strength. (From Bilgin, N. et al., *Int. J. Rock Mech. Min. Sci.*, 73, 123–129, 2015.)

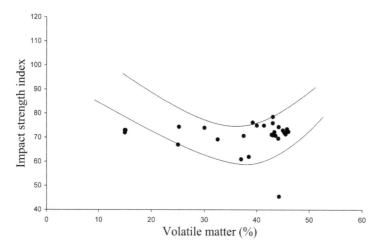

FIGURE 5.9
Relationship between volatile matter and impact strength index. (From Bilgin, N. et al., *Int. J. Rock Mech. Min. Sci.*, 73, 123–129, 2015.)

After Su et al. (2003), for Zonguldak Coal Field, ISI was related to volatile matter, carbon content and Hardgrove index, as given in Figures 5.10 through 5.12. Although the values in these figures are different in magnitude than obtained for Soma-Imbat Mine, they show the same trends as previous ones.

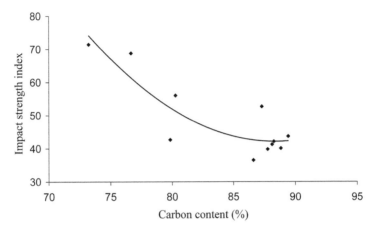

FIGURE 5.10
Relationship between impact strength index and carbon content for Turkish hard coals from Zonguldak. (From Su, O. et al., The grindability and impact strength index of Zonguldak bituminous coals, *Mineral Processing in the 21th Century: Proceedings of X Balkan Mineral Processing Congress,* Varna, Bulgaria, June 15–20, 2003, 275–279, 2003. Used with kind permission of Okan Su.)

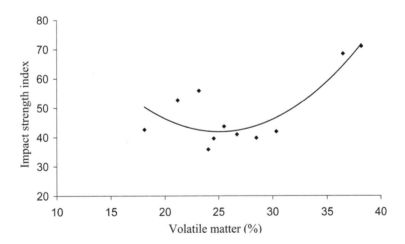

FIGURE 5.11
Relationship between impact strength index and volatile matter for Turkish hard coals from Zonguldak. (From Su, O. et al., The grindability and impact strength index of Zonguldak bituminous coals, *Mineral Processing in the 21th Century: Proceedings of X Balkan Mineral Processing Congress,* Varna, Bulgaria, June 15–20, 2003, 275–279, 2003. Used with kind permission of Okan Su.)

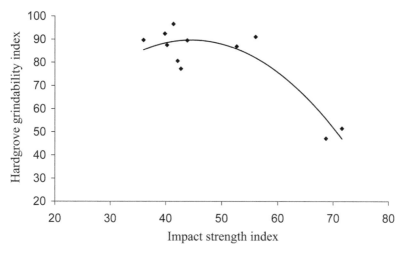

FIGURE 5.12
Relationship between impact strength index and Hardgrove grindability index for Turkish coals. (From Su, O. et al., The grindability and impact strength index of Zonguldak bituminous coals, *Mineral Processing in the 21th Century: Proceedings of X Balkan Mineral Processing Congress*, Varna, Bulgaria, June 15–20, 2003, 275–279, 2003. Used with kind permission of Okan Su.)

5.10 Cone Indenter Hardness

The cone indenter was designed at the Mining Research and Development Establishment of the previous National Coal Board (NCB) of England to determine the resistance of rock and coal to indentation by a sharp tungsten carbide cone with 60° tip angle (NCB 1977, Szlavin 1974). It was designed to determine the hardness of small fragments of coal by measuring its resistance to indentation by a hardened tungsten carbide cone. A typical example of a NCB cone indenter device is seen in Figure 5.13. A specimen about 12 mm × 12 mm × 6 mm in size is placed on the steel strip and the cone is lowered by turning the micrometer under 40 N force. Displacement between the first and the second advancement is read (M1 and M2). The deflection of the thin spring bond is measured by the gauge and is directly related to the force on the specimen. The correlation between the standard cone indenter hardness (CIH) and σ_c has been determined by the NCB and is given in Equation 5.6.

$$\sigma_c = CIH \cdot 24.8 \tag{5.6}$$

FIGURE 5.13
Cone indenter device of NCB.

Equation 5.6 should be used in caution since, for example, for CIH, a value of 2.5 gives a compressive strength value of 236 kg/cm² for coals from Soma Turkey, as seen in Figure 5.14, and Equation 5.6 gives a value of 629 kg/cm² for the same CIH value of coals from England. This difference may come from the petrological differences of the coals from two different regions.

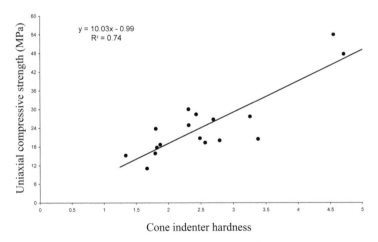

FIGURE 5.14
Relationship between cone indenter hardness and uniaxial compressive strength. (From Bilgin, N. et al., *Mechanical Excavation in Mining and Civil Industries*, CRC Press/Taylor & Francis Group, Boca Raton, FL, 2014a.)

5.11 Shore Scleroscope Hardness

Shore scleroscope hardness is one of the simplest methods given to determine the surface hardness of the tested material. It is determined by the rebound height of a diamond or tungsten-carbide-tipped hammer dropped onto a horizontal smooth surface. In Shore scleroscope test, a diamond tip is dropped from a fixed height into the rock specimen. The hammer then rebounds, but not to its original height because some of the energy in the falling tip is dissipated in producing an indentation. The instrument used is supplied in two models designated model C and model D. Model C-2 consists of a vertically disposed barrel containing a glass tube which is graded from 0 to 140, as seen in Figure 5.15 (Bilgin et al. 2014b). A diamond tip is dropped from a specified height and rebounds within the glass tube. According to the suggested methods published by the ISRM (2007), a test specimen having a minimum surface area of 10 cm^2 and a minimum thickness of 1 cm is necessary. Measurement points should have at least 5 mm distance from each other and only one test must be carried out at the same spot. The minimum

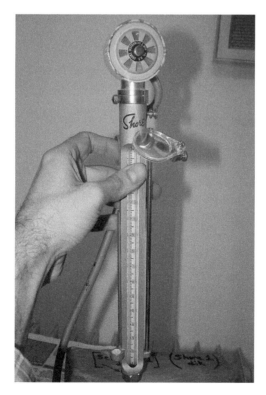

FIGURE 5.15
Shore scleroscope hardness test device.

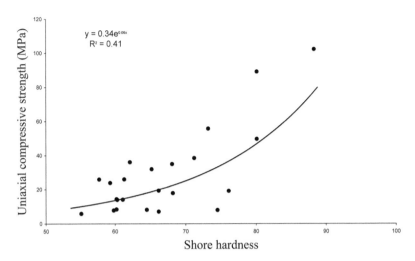

FIGURE 5.16
Relationship between Shore scleroscope hardness and uniaxial compressive strength for coal samples taken from Soma Mine, Turkey. (From Bilgin, N. et al., The cuttability and cavability of thick coal seam in Soma-Eynez coal field IR-75153 of TKI operated by IMBAT AS, Project report, Istanbul Technical University, 59, 2011.)

number of tests for each sample is recommended to be 20 for statistical reliability (ISRM 2007), explained in detail in Tumac et al. (2007) and Altindag and Guney (2006).

An extensive set of mechanical tests on coal (lignite) specimens was carried out for Imbat Mine (Soma, Turkey) (Bilgin et al. 2011). The relationship between Shore scleroscope hardness and compressive strength shown in Figure 5.16 is extracted from this report. Although, as seen in this figure, the Shore hardness values changed between 15 and 110, in the other two examples from Amasra hard coal field, Turkey, the mean values changed between 24 and 85 (Bilgin et al. 2010, 2014b).

5.12 Hardgrove Grindability Index

The Hardgrove grindability index (HGI) was developed in the 1930s as an empirical test to indicate how difficult it would be to grind a specific coal to the particle size necessary for effective combustion in a pulverized coal fired boiler. The test is based on Rittinger's theory that "the work done in grinding is proportional to the new surface produced."

The index varies from 20 to 110, with a lower HGI indicating a coal is harder to grind. The test is conducted on a standardized laboratory scale ball-and-race mill and is defined in British Standard 1016-112: (1995) or ASTM D409 (2016) standard. A strong hard coal will often have a high rank and be difficult to reduce in size, a weak, soft coal of lower rank will be easier to grind, but very low rank coals can also be difficult to reduce in size. For coal, HGI correlates to uniaxial compressive and tensile strength measurements, which roughly correlate with coal rank (Williams et al. 2015).

Power plants burning ground coal usually require values of HGI > 60, coal with a value of HGI < 50 is considered heavily grindable in North America. Australian coal was in terms of HGI grindability divided into six groups (Agus and Waters 1971). The coal types with grindability value below 40 HGI are classified as very hard, coals with HGI value from 40 to 60 are classified as hard, the coal with HGI value from 60 to 80 as medium hard, from 80 to 100 HGI as soft, and with HGI from 100 to 120 as very soft, and coal with the grindability value HGI above 120 is classified as extremely soft (this group, however, includes only some brown types of coal).

Su et al. (2003) investigated the effect of different parameters on the HGI of the samples taken from Zonguldak, Turkey Hard Coal Field. They found similar trends as given by Szwilski (1985), as illustrated in Figures 5.17 and 5.18.

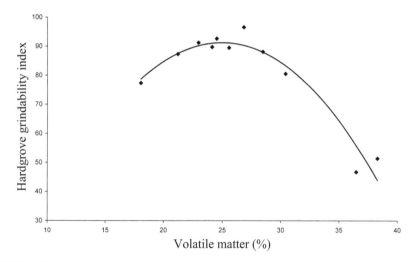

FIGURE 5.17
Relationship between Hardgrove grindability index and volatile matter for Turkish coals. (From Su, O. et al., The grindability and impact strength index of Zonguldak bituminous coals, *Mineral Processing in the 21th Century: Proceedings of X Balkan Mineral Processing Congress*, Varna, Bulgaria, June 15–20, 2003, 275–279, 2003. Used with kind permission of Okan Su.)

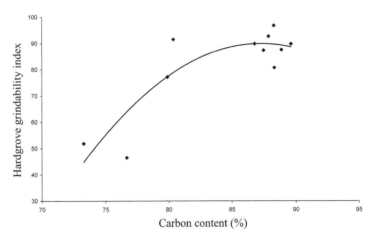

FIGURE 5.18
Relationship between Hardgrove grindability index and carbon content for Turkish coals. (From Su, O. et al., The grindability and impact strength index of Zonguldak bituminous coals, *Mineral Processing in the 21th Century: Proceedings of X Balkan Mineral Processing Congress,* Varna, Bulgaria, June 15–20, 2003, 275–279, 2003. Used with kind permission of Okan Su.)

5.13 Conclusive Remarks

The Physical and mechanical properties of coal are of great importance in coal mining engineering.

The density of coal shows a notable variation with the rank of carbon content. Coal bed methane is one of the unconventional energy sources, where many countries have started to explore seriously. Determination of coal cleat porosity and permeability are necessary for prediction of methane production rates in coalbed methane operations.

The uniaxial compressive strength of coal is an important mechanical property of coal, which is difficult to investigate. The following factors may effect directly the compressive strength of coal and they should be taken into consideration for any engineering design: (a) the rank or petrology of coal, (b) plane of weaknesses, bedding planes, and cleats, (c) specimen size, and (d) confining stress. It is shown that the uniaxial compressive strength of coal may be predicted from point load, cone indenter hardness, Shore scleroscope hardness, and Schmidt hammer and impact strength harness tests.

It is essential to understand the factors effecting the tensile strength of coal since, as it will be explained within this book, tensile breakage plays an important role in coal cutting mechanics.

Young's modulus of coal is a critical parameter in design of coal mining and coalbed methane development, coal pillar stability, coal outburst, and the other related engineering design topics.

Poisson's ratio of coal is also an important parameter of coal, which is mostly used to predict the sonic velocities of coal.

The HGI is an empirical test to indicate how difficult it would be to grind a specific coal to the particle size necessary for effective combustion in a pulverized coal fired boilers. The coal types with value below 40 HGI are classified as very hard and the coal with HGI value from 60 to 80 as medium hard.

References

ASTM D167. 2012. Standard Test Method for Apparent and True Specific Gravity and Porosity of Lump Coke.

ASTM D3967. 2016. Standard Test Method for Splitting Tensile Strength of Intact Rock Core Specimens.

ASTM D409. 2016. Standard Test Method for Grindability of Coal by the Hardgrove-Machine Method.

Agus, F. and Waters, P. L. 1971. Determination of the grindability of coal, shale and other minerals by a modified hardgrove machine method. *Fuel,* 50:405–431.

Altindag, R. and Guney, A. 2006. ISRM Suggested Method for determining the Shore Hardness value for rock. *International Journal of Mining Science and Technology,* 43(1):19–22.

Al-Jubori, A., Johnston, S., Boyer, C., Lambert, S. W., Bustos, O. A., Land, S., Jack, C., Pashin, J. C., and Wray, A. 2009. Coalbed methane: Clean energy for the world. *Oilfield Review,* 21(2):4–14.

Anand, V. and Gri, M. 2015. *Evaluation and Correlation of Some Properties of Coal.* National Institute of Technology, Rourkela, India, 55 p.

Arioglu, E. and Tokgoz, N. 2011. *Hard Rock Mass and Coal Strength.* Evrim publisher, Istanbul, Turkey, 280 p. (in Turkish).

Atkinson, R. H. and Ko, H. Y. 1977. Strength characteristics of US Coals. *American Rock Mechanics Association, the 18th U.S. Symposium on Rock Mechanics (USRMS),* June 22–24, Golden, Colorado, pp. 283/1–6.

Bieniawski, Z. T. 1984. *Rock Mechanics Design in Mining and Tunneling.* Balkema Publications, Rotterdam, the Netherlands.

Bilgin, N., Akgun, S.I., and Shahriar, K. 1988. The classification of coal seams in Amasra coalfield according to their mechanical strength. *The Sixth Coal Congress of Turkiye,* May, Zonguldak, Turkiye, in Turkish, pp. 13–28.

Bilgin, N., Balci, C., Copur, H., Tumac, D., and Avunduk, E. 2015. Cuttability of coal from the Soma coalfield in Turkey. *International Journal of Rock Mechanics and Mining Sciences,* 73:123–129.

Bilgin, N., Copur, H., and Balci, C. 2014a. *Mechanical Excavation in Mining and Civil Industries.* CRC Press/Taylor & Francis Group, Boca Raton, FL.

Bilgin, N., Copur, H., Balci, C., Avunduk, E., and Tumac, D. 2011. The cuttability and cavability of thick coal seam in Soma-Eynez coal field IR-75153 of TKI operated by IMBAT AS. Project report, Istanbul Technical University, Turkey 59 p.

Bilgin, N., Copur, H., Balci, C., Tumac, D., Avunduk, E., and Comakl, R. 2014b. Cuttability studies for a Coal Mine in Zonguldak. Project report, Istanbul Technical University, Turkey, 49 p.

Bilgin, N., Phillips, H. R., and Yavuz, N. 1992. The cuttability classification of coal seams and an example to a mechanical application in ELİ Darkale Coal Mine. *Proceedings of the 8th Coal Congress of Turkey*, Zonguldak, Turkey, pp. 31–51.

Bilgin, N., Temizyurek, I., Copur, H., Balci, C., and Tumac, D. 2010. Cuttability characteristics of TTK Amasra thick seam and some comments on mechanized excavation. *Proceedings of the 17th Coal Congress of Turkey*. June 2–4, Zonguldak, Turkey, pp. 217–229 (in Turkish).

Broch, G. and Franklin, J. A. 1972. The point load strength test. *International Journal of Rock Mechanics and Mineral Science and Geomechanics Abstracts*, 9:669–697.

Brook, N. 1977. The use of irregular specimens for rock strength tests. *International Journal of Rock Mechanics and Mineral Science and Geomechanics Abstracts*, 14:193–202.

Brook, N. 1985. The equivalent core diameter method of size and shape correction in point load testing. *International Journal of Rock Mechanics and Mineral Science and Geomechanics Abstracts*, 22:61–70.

BS 1016-112. 1995. Methods for analysis and testing of coal and coke. Determination of Hardgrove Grindability Index of Hard Coal.

Copur, H. and Balci, C. 2010. Roadheader selection and cuttability of lignite and measure stones for Polyak-Eynez Enerji Madencilik A.Ş. Kınık Elmadere Köyü Coal Field. Project report, Istanbul Technical University, Turkey, 60 p.

Durucan, S., Mustafa, M., Syed, A., Shi, J. Q., and Korre, A. 2013. Two phase relative permeability of gas and water in coal for enhanced coalbed methane recovery and CO_2 storage. *Energy Procedia*, 37:6730–6737.

Evans, I. and Pomeroy, C. D. 1966. *The Strength, Fracture and Workability of Coal*. Pergamon Press Ltd, London, UK.

Gaddy, F. L. 1956. A study of the ultimate strength of coal as related to the absolute size of the cubical specimens tested. *Bulletin 112 Virginia Engineering Experiment Station*, Virginia Polytechnic Institute, Blacksburg, VA.

Gash, B. W., Volz, R. F., Potter, G., John, M., and Corgan, J. M. 1992. The Effects of cleat orientation and confining pressure on cleat porosity, permeability and relative permeability in coal. *Proceedings of SCA Conference*, paper number 9224, pp. 1–26.

Gonzatti, C., Zorzi, L., Agostini, I. M., Fiorentini, J. A., Viero, A. P., and Philipp, R. P. 2014. In situ strength of coal bed based on the size effect study on the uniaxial compressive strength. *International Journal of Mining Science and Technology*, 24:747–754.

Gosh, N. 2011. Permeability of Indian coal. BSc diss., Department of Mining Engineering National Institute of Technology Rourkela, India, 47 p.

Holland, C. T. 1964. *The Strength of Coal in Mine Pillars*. School of Mines, West Virginia University, Morgantown, WV, 450–466.

ISRM. 2007. The complete ISRM suggested methods for rock characterization, testing and monitoring 1974–2006. In eds. R. Ulusay and J.A. Hudson, *Compilation Arranged by the ISRM*. Turkish National Group, Ankara, Turkey, 628.

Krey, T. C. 1963. Chanel waves a tool of applied geophysics in coal mining. *Geophysics*, 28:701–714.

Kumar, H., Mishra, S., and Mishra, M. K. 2015. Experimental evaluation of geo-mechanical properties of coal using sonic wave velocity. *International Conference on Advances in Agricultural, Biological & Environmental Sciences (AABES-2015)*, July 22–23, 2015, London, UK.

Morcote, A., Mavko, G., and Prasad, M. 2010. Dynamic elastic properties of coal. *Geophysics*, 75(6):E227–E234.

Mathey, M. 2015. *Investigation into the Mechanism of Strength and Failure in Squat Coal Pillars in South Africa*. University of Witwatersrand, Johannesburg, South Africa, 202 p.

Mitra, A., Harpalani, S., and Liu, S. 2012. Laboratory measurement and modeling of coal permeability with continued methane production: Part 1—Laboratory results. *Fuel*, 94:110–116.

NCB. 1977. *NCB Cone Indenter. MRDE Handbook No. 5. Staffordshire: Mining Research and Development Establishment*. National Coal Board, Bretby, UK.

Okubo, S., Fukui, K., and Qi, Q. X. 2006. Uniaxial compression and tension tests of anthracite and loading rate dependence of peak strength. *International Journal of Coal Geology*, 68(3–4):196–204.

Pan, J., Meng, Z., Hou, Q., Ju, Y., and Cao, Y. 2013. Coal strength and Young's modulus related to coal rank, compressional velocity and maceral composition. *Journal of Structural Geology*, 54:129–135.

Quinteiro, C., Galvin, J., and Salamon, N. S. W. 1995. Coal pillar yield mechanics, *ISRM 8th conference*, 25–29 September, Tokyo, pp. 1359–1362.

Ramandi, H. L., Mostaghimi, P., Armstrong, R. T., Saadatfar, M, and Pinczewski, W. V. 2016. Porosity and permeability characterization of coal: A micro-computed tomography study. *International Journal of Coal Geology*, 154–155:57–68.

Roxborough, F. F. and Hagen, P. 2010. Elements of machine mining, module reader. *Mining Education Australia (MEA)*. Minerals Council of Australia, 88 p.

Roxborough, F. F., King, P., and Pendroncelli, E. J. 1981. Tests on the cutting performance of a continuous miner. *Journal of South African Institute of Mining and Metallurgy*, 81:9–26. January, 9–25.

Salamon, M. D. G. and Munro, A. H. 1967. A study of the strength of coal pillars. *Journal of the Southern African Institute of Mining and Metallurgy*, 68(2):55–67.

Speight, J. 2015. *Handbook of Coal Analysis*. Wiley, New York.

Su, O., Toroglu, I., and Akcın, A. A. 2003. The grindability and impact strength index of Zonguldak bituminous coals. *Mineral Processing in the 21th Century: Proceedings of X Balkan Mineral Processing Congress*, Varna, Bulgaria, June 15–20, 2003, pp. 275–279.

Szabo, T. L. 1981. A representative Poisson's ratio for coal. *International Journal of Rock Mechanics and Mineral Science and Geomechanics Abstracts*, 18:531–533.

Szlavin, J. 1974. Relationships between some physical properties of rock determined by laboratory tests. *International Journal of Rock Mechanics and Mineral Science and Geomechanics Abstracts*, 11:57–66.

Szwilski, A. B. 1984. Determination of the anisotropic elastic moduli of coal. *International Journal of Rock Mechanics and Mineral Science and Geomechanics Abstracts*, 21(1):3–12.

Szwilski, A. B. 1985. Relation between the structural and physical properties of coal. *Mining Science and Technology*, 2:181–189.

Townsend, J. M., Jennings, W. C., Haycocks, C., Niall, G. M., and Johnson, L. P. 1977. A relationship between the ultimate compressive strength of cubes and cylinders for coal specimens. *American Rock Mechanics Association, the 18th U.S. Symposium on Rock Mechanics (USRMS)*, June, 22–24, Golden, Colorado, pp. 4A6-1–4A6-6.

Tumac, D., Bilgin, N., Feridunoglu, C., and Ergin, H. 2007. Estimation of rock cutta-bility from shore hardness and compressive strength properties. *Rock Mechanics and Rock Engineering*, 40(5):477–490.

Udd, J. E., Hudson, K., and Templeton, D. 1980. Compressive strength of a bitumi-nous coking coal. *American Rock Mechanics Association, the 21st U.S. Symposium on Rock Mechanics (USRMS)*, May, 27–30, Rolla, Missouri, pp. 468–475.

Williams, O., Eastwick, C., Kingman, S., Giddings, A., Lormor, S., and Lestera, E. 2015. Investigation into the applicability of Bond Work Index (BWI) and Hardgrove Grindability Index (HGI) tests for several biomasses compared to Colombian La Loma coal. *Fuel*, 158:379–387.

Yazici, S., Acaroglu, O., Arapoglu, B., Bilgin, N., and Eskikaya, S. 1998. Investigation into of using continuous miners in an opencast coal mine. *Proceedings of the 8th Coal Congress of Turkey*, Zonguldak, Turkey. pp. 11–20.

Zhang, J., Feng, Q., Zhang, X., Wen, S., and Zhai, Y. 2015. Relative permeability of coal: A review. *Transport in Porous Media*, 106(3):563–594.

Zhang, Q. B. and Zhao, J. 2014. A review of dynamic experimental techniques and mechanical behaviour of rock materials. *Rock Mechanics and Rock Engineering*, 47(4):1411–1478.

Zhao, Y., Liu, S., Jiang, Y., Wang, K., and Huang, Y. 2016. Dynamic tensile strength of coal under dry and saturated conditions. *Rock Mechanics and Rock Engineering*, 49(5):1709–1720.

Zhao, Y., Zhao, G. F., Jiang, Y., Elsworth, D., and Huang, Y. 2014. Effects of bedding on the dynamic indirect tensile strength of coal: Laboratory experiments and numerical simulation. *International Journal of Coal Geology*, 132:81–93.

Zhou, Y. X., Xia, K., Li, X. B., Li, H. B., Ma, G. W., Zhao, J., Zhou, Z. L., and Dai, F. 2012. Suggested methods for determining the dynamic strength parameters and mode-I fracture toughness of rock materials. *International Journal of Rock Mechanics and Mining Sciences*, 49:105–112.

6

Schmidt Hammer Hardness, In-Situ Strength of Coal, Cleats

6.1 Introduction

This chapter is a summary of a paper recently published by the authors in Bilgin et al. (2016). However, it is extended in a way that more information on the subject is added to make the subject more comprehensive.

A study to see the effect of cleats and to develop an in-situ strength and excavatability classification system for coal seams based on Schmidt hammer tests was carried out by the authors for Imbat Coal (Lignite) Mine located in Soma Coal Basin of Turkey. Field tests with a N-type Schmidt hammer were realized covering a total number of around 1,350 tests. Coal samples were also collected to determine some of their physical and mechanical properties in the laboratory including compressive strength, dynamic, and static elasticity modulus. Cleat spacings were also measured in the laboratory with a caliper by visually examining the coal samples. A statistical analysis was carried out to see the effect of cleat spacing (frequency) on the in-situ strength and excavatability of the coal seam. A strength reduction factor as previously defined by another researcher group for rock discontinuities was used to see the effect of cleats on Schmidt hammer rebound values (Young and Fowell 1978). The strength values obtained in the laboratory were correlated with Schmidt hammer rebound values (mean value of ten readings, the highest value, mean of the highest 2, 3, 4, and 5 Schmidt hammer readings) to develop a methodology for using Schmidt hammer rebound value in excavatability/workability classification of coal seams.

The results indicate that the mean of the highest three Schmidt hammer rebound values within a set of ten readings on a certain location of the coal seam gives the highest correlations with uniaxial compressive strength and Young's modulus in all cases and can be reliably used for excavatability classification of the coal seams.

6.2 General Information on Schmidt Hammer and Standards Used

Schmidt hammer is a portable device developed in 1948 to measure the surface hardness of concrete, and it has been later used extensively to test in-situ strength of rock formations and coal. The main working principle is realized by transforming potential energy to kinetic energy by means of a spring and hammer within the testing device. Mainly two standards are used for Schmidt hammer tests, American Society for Testing and Materials (ASTM) D5873-14 (2015) and ISRM standard reported by Ulusay (2015).

The standard suggested in the "ISRM suggested methods for rock characterization, testing and monitoring" given in Ulusay (2015) is based on detailed work carried out by Basu and Aydin (2004), Aydin and Basu (2005), and Aydin (2009). In their studies, the influence of hammer type, direction of hammer impact relative to the horizontal or vertical plane, specimen requirements, weathering, moisture content and testing data gathering/reduction, and analysis procedures were considered in detail. In this standard, it was pointed out that L- and N-type hammers, with respective impact energies of 0.735 Nm and 2.207 Nm, should be used with caution when the uniaxial compressive strength of the rock material is outside the range between 20 MPa and 150 MPa, where sensitivity decreases and data scatter increases. The N-type hammer is less sensitive to surface irregularities and should be preferred in field applications, while the L-type hammer has greater sensitivity in the lower range and gives better results when testing weak, porous, and weathered rocks in the laboratory (Ozbek 2009). For data gathering, 20 rebound values should be recorded from single impacts separated by at least a plunger diameter (to be adjusted according to the extent of impact crater and radial cracks). On the other hand, the test may be stopped when any ten subsequent readings differ only by four (corresponding to Schmidt hammer repeatability range of ±2). In rocks such as coal, shale, and slate, testing over lamination walls may produce a narrow range of rebound values due to their uniform and smooth nature, but also significantly low values due to these interfaces.

ASTM standard for Schmidt hammer (ASTM D5873-14, 2015) points out that the test is best suited for rock material with uniaxial compressive strengths ranging between approximately 1 MPa and 100 MPa and the rebound hardness value can serve in a variety of engineering applications that require characterization of rock material. These applications include, for example, prediction of penetration rates for tunnel boring machines, determination of rock quality for construction purposes, grouping of test specimens, and prediction of hydraulic erodibility of rock. Rock at 0°C or less may exhibit very high rebound values. Temperature of the rebound hammer itself may affect the rebound value. The hammer and materials to be tested should be at the same temperature. For readings to be compared, the direction of impact must be the same. Different instruments of the same nominal design may give rebound values differing from one to three units and, therefore, tests should be performed with the same instrument to obtain comparable results. If more than one instrument is to be used, a sufficient number of tests must be performed on typical rock surfaces to determine the magnitude of differences to be expected in the readings of different instruments. Samples can be drill cores having a size of Core of 54.7 mm in diameter (NX) or larger, rock blocks, or in-situ rock surfaces, such as tunnel or gallery walls. The test surface of all specimens, either in the laboratory or in the field, should be smooth to the touch or free of joints, fractures, or other obvious localized discontinuities to a depth of at least 6 cm. In-situ rock surface should be flat and free of surface grit over the area covered by the plunger. If the surface of the test area is heavily textured, it should be ground with an abrasive stone to smooth the surface. According to ASTM D5873-14 (2015), ten representative locations on the specimens should be selected. Test locations should be separated by at least the diameter of the plunger and only one test/reading may be taken at any one point. The average of the ten readings should be obtained for each specimen to the nearest whole number.

6.3 A Brief Summary of Previous Research Studies on Schmidt Hammer

The Schmidt hammer test has been widely used in civil engineering to define surface strength of concrete, however, its use in rock mechanics and mining applications have also been the subject of several research works in the past. Roxborough and Whittaker (1964, 1965) and Kidybinski (1968) were some of the first users of Schmidt hammer to test the strength of coal and roof strata in coal mining. Carter and Sneddon (1977) compared Schmidt hammer values with the point load and uniaxial compressive strength of rocks. The research published by Young and Fowell (1978) described the use of a Schmidt hammer with a grid on a tunnel face to determine the Schmidt hammer value reduction index. The index was shown to be related to excavation rates of roadheaders, with increases in these rates being attributed to the fractured state of the rock mass. The authors considered that the relationships established for mudstone should also be applicable to most sedimentary rocks, though the values of intercepts and gradients would vary from site to site. Poole and Farmer (1978) used Schmidt hammer values in predicting performance of roadheaders. Shorey et al. (1984) performed Schmidt hammer tests to determine in-situ strength of coal seams. Haramy and DeMarco (1995) found statistically reliable correlations between uniaxial compressive strength and Schmidt hammer rebound values of coal. Ghose and Chakraborti (1986) correlated Schmidt hammer values with uniaxial compressive strength, tensile strength, impact strength index, and cone indenter values. Sachpazis (1990) emphasized the importance of lithological classification of rock formations to find reliable statistical relations between rock properties and Schmidt hammer rebound values. Ayday and Goktan (1992) and Goktan and Ayday (1993) did comprehensive studies on Schmidt hammer and correlated N- and L-type Schmidt hammer test results. Poole and Farmer (1978, 1980), Katz et al. (2000), Kahraman (2001a, 2001b), Kahraman et al. (2002), Yilmaz and Sendir (2002), Yasar and Erdogan (2004), Dincer et al. (2004), Fener et al. (2005), Buyuksagis and Goktan (2007), Cobanoglu and Celik (2008), Yagiz (2009), Demirdag et al. (2009), Niedzielski et al. (2009), Ozbek (2009), Sharma et al. (2011), Sengun et al. (2011), Bruno et al. (2013), Minaeian and Ahangari (2013), Saptono et al. (2013), Karaman and Kesimal (2015), Tandon and Gupta (2015), and Wang et al. (2017) are among the researchers who did immense research works to correlate rock properties with Schmidt hammer rebound values.

Bilgin et al. (2002) found a good correlation between Schmidt hammer rebound values and net breaking rate of impact hammers used in Istanbul Metro drivages. Ozkan and Bilim (2008) and Bilim and Ozkan (2008) investigated the change of Schmidt hammer rebound values with changing coal roof strata conditions related to mining activities. According to Vakili and Hebblewhite (2010), basic parameters affecting excavatability/workability and cavability of a coal seam were thickness of coal seam, strength characteristics and elasticity modulus of coal, overburden, and spacing of face and butt cleats.

Greco and Sorriso-Valvo (2005) carried out an intensive research program in order to investigate how the apparent separation of jointing varies according to the distance from faults, and how the mechanical properties of rock masses depend on this distance and jointing density. A number of regression analyses were performed on the variables "s" (apparent joint separation), "d" (distance from major fault), and "SH" (rebound value of Schmidt hammer). The variables were measured at 380 stations distributed over a wide study area located in Calabria, Southern Italy. They concluded that there was a relationship between the "SH" values and the "d" values. Finally, the relationship between "SH" and "s" remained problematic.

Tumac (2015) in his study on "Predicting the Performance of Large Diameter Circular Saws Based on Schmidt Hammer and Other Properties for Some Turkish Carbonate Rocks" used deformation coefficient (K) defined by Fowell and McFeat-Smith (1976). According to these authors, a plot of the hardness against test number at the same point showed that the readings increased initially and then maintained a constant level after 20 tests, with only a small variation. Deformation coefficient (K) could be calculated as the percentage difference between the initial and the final constant value. The results obtained from this approach were correlated with the performance of roadheaders. Tumac (2015) reported that the performance of large diameter circular saws could also be predicted by (K) values.

In thick coal seam mining where sub-level caving method is used, it is vital to know the changing strength characteristics of different levels of coal seams, which would determine the excavatability/workability of the seam for planning of the mine. The Schmidt hammer seems to be one of the best methods to be used for this purpose. However, according to the best knowledge of the authors of this study, there is not any information in the published literature on how the Schmidt hammer rebound value is affected by the existence and spacing of cleats.

The main objective of this chapter is to see whether it is possible to define a Schmidt hammer strength reduction factor related to cleat occurrence and cleat spacing, and a methodology of using in-situ Schmidt hammer rebound values to correlate with laboratory coal strength values, and thus to develop a new excavatability/workability classification system for coals. For this purpose, an underground coal mine having around 30 m of seam thickness located in Soma Coal Basin of Turkey was selected for sampling and field tests with an N-type Schmidt hammer. Around 1,350 Schmidt hammer test readings corresponding to 134 different locations in the mine were carried out and coal samples were collected to determine some of their physical and mechanical properties to identify the variability of coal strength along the thickness of the coal seam. Cleat spacings were measured in the laboratory by examining visually the samples obtained from the field. A statistical analysis was carried out to see the effect of cleat spacing on the in-situ strength and workability of the coal. A strength reduction factor as defined by Young and Fowell (1978) for rock discontinuities is used in this study to see the effect of cleats on Schmidt hammer values. Some strength tests such as uniaxial compressive strength, dynamic, and static elasticity modulus were also carried out in the laboratory and the results were correlated with Schmidt hammer rebound values performed at different levels in the coal seam (mean value of ten readings, the highest value, mean of the highest 2, 3, 4, and 5 values of the Schmidt hammer readings) to develop a methodology for using Schmidt hammer rebound value in excavability classification of coal seams.

6.4 Cleats and Their Effect on the Schmidt Hammer Rebound Values

Natural vertical fracture systems in bituminous coalbeds are called "cleat." Cleat orientation commonly controls the direction of mining with major development paralleling the face cleat. Previous researchers categorized the origin of cleat as endogenetic relating the origin of cleat to compaction and coalification and exogenetic relating the origin of cleat to tectonic forces. Endogenetic cleat is formed during the process of physical changes in the properties of coal during the metamorphic process. Coal matter undergoes density changes and a decrease in its volume. These processes are associated with the changes in the internal stress system, compaction and desiccation, and the formation of cleat planes. Exogenetic cleat is formed as a result of the external stresses acting on the coal seam, which include tectonic stresses, fluid pressure changes, folding, and development of tensile stresses that coal seam is subjected during various time periods. Endogenetic cleats are normal to the bedding plane of coal and generally occur in pairs. There are at least two sets of near perpendicular fractures that intersect the coal to form an interconnected network throughout a coalbed. These two fracture systems are known as face and butt cleats. The shorter butt cleat normally terminates at a face cleat, which is the prominent type of cleat (McCulloch et al. 1974, Dawson and Esterle 2010). Face cleats are formed as extension fractures during structural deformation, and butt cleats as release fractures during erosion and uplift. Face and butt cleats are illustrated in Figures 6.1 and 6.2.

Directional permeability of coal is directly related to cleats. Holes drilled perpendicular to the face cleats yield from 2.5 times to 10 times the amount of released gas as compared with holes drilled perpendicular to the butt cleats (McCulloch et al. 1974). The cleat directions, especially in bituminous coals, are important in establishing preferred directions of mine development. The coal tends to break along its natural fracture systems. Therefore, it is much easier to mine parallel to the cleat directions than at an angle to the cleats. Evans and Pomeroy (1966) investigated the effect of cleat orientation on the workability of coal and emphasized that cleat spacing, cleat orientation, and the strength of coal had major effects on the cuttability of

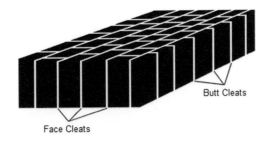

FIGURE 6.1
Illustration of face and butt cleats. (From Dawson, G.K.W. and Esterle, J.S., *Int. J. Coal. Geol.*, 82, 213–328, 2010.)

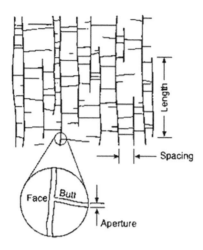

FIGURE 6.2
Illustration of coal cleat geometries with cleat-trace patterns in plain view. (From Laubacha, S.E. et al., *Int. J. Coal. Geol.*, 35, 175–207, 1998.)

coal seams. Vakili and Hebblewhite (2010) demonstrated that strength and number of fractures had an effect on the cavability of thick coal seams. The efficient use of coal excavating machines like ploughs or shearers depend mainly on coal strength.

Cleat spacing, length, height, connectivity, aperture size, degree of mineral infilling, type of mineral infilling, and angle of orientation all affect the ability of fluids to travel through coal and, hence, the rate and volume of methane which can be extracted. Thus, a clear understanding of cleats is essential for the success of any coal seam gas venture and coal workability (Dawson and Esterle 2010).

An intensive review of cleats is made by Laubacha et al. (1998) indicating that fractures occurred in nearly all coalbeds and could exert fundamental control on coal stability, minability, and fluid flow. It is therefore not surprising that coal fractures have been investigated since the early days of coal mining. Although fractures in coal are relatively unimportant in strip mining, their significance in efficient design and safety of underground coal mines has continued to command the attention of the mining industry. Measurements on long mine highwalls showed that spacing between individual cleats of similar size ranged from microns to more than a meter. Yet within a given bed, average spacing of fractures of similar size was remarkably uniform over distances of hundreds of meters (Laubacha et al. 1998).

The angle between the face and butt cleats is around 90°. The spacing between cleats varies according to factors such as the coal maturity, the mineral matter, and the carbon content, but normally is less than 25 mm. Coals with bright lithotype layers, with a high percentage of vitrinite macerals, have a greater frequency of cleats than dull coals. Common understanding is that cleats are formed due to the effects of the intrinsic tensile force, fluid pressure, and tectonic stress. The intrinsic tensile force arises from matrix shrinkage of coal, and the fluid pressure arises from hydrocarbons and other fluids within the coal. These two factors are considered to be the reasons for endogenetic cleat formation. On the other hand, the tectonic stress is regarded as extrinsic to cleat formation and is the major factor that controls the geometric pattern of cleats. Face cleats extend in the direction of maximum in-situ stress, and butt cleats extend in the direction of minimum in-situ stress which occurs at the time of their formation. Face cleats are developed perpendicular to bedding planes, and butt cleats are developed parallel to bedding planes (McCulloch et al. 1974, Laubacha et al. 1998, Dawson and Esterle 2010). Cleat frequency or cleat spacing is a characteristic of the coal basin development. For example, in some regions in Australia, the distance between cleats varies between 10 mm and 25 mm and the aperture between 0.1 mm and 0.2 mm.

One of the most important parameters affecting rock mass strength is the number of joints or fractures. Young and Fowell (1978) observed that Schmidt hammer rebound value decreased with increasing number of joints and defined Schmidt hammer strength reduction factor (SHRF) as the percentage of the difference between the maximum and mean values of Schmidt hammer. With a similar way of thinking, an attempt to see the effect of cleat spacing on Schmidt hammer rebound values and to obtain Schmidt hammer strength reduction factor if it exists, fifty coal samples were collected in this study from the locations where Schmidt hammer readings were taken. During the preparation of samples for mechanical tests, cleats were visually counted and their average spacings were estimated in some samples where the cleats were apparent. Typical samples with cleats obtained from the mine are shown in Figure 6.3. A statistical analysis was carried out to see the effect of the spacing (frequency) of the cleats on the in-situ strength and excavatability/workability of the coal.

Statistical percentile analysis was carried out to see the effect of cleat spacings on Schmidt hammer values. In this analysis, strength reduction factor (SHRF), as defined by Young and Fowell (1978), is calculated by Equation 6.1.

$$SHRF = \left[\frac{SH_{MAX} - SH_{MEAN}}{SH_{MEAN}} \right] \cdot 100, \quad (\%) \qquad (6.1)$$

where, SHRF is Schmidt hammer strength reduction factor, SH_{MAX} is maximum Schmidt hammer rebound value (in one set of reading on a

FIGURE 6.3
Typical samples with cleats obtained from the Imbat Lignite Mine.

location), and SH_{MEAN} is mean Schmidt hammer rebound value of a set of readings at the same location. In this study, the mean values of Schmidt hammer reduction factors ($SHRF_1$, $SHRF_2$, $SHRF_3$, $SHRF_4$, and $SHRF_5$) are estimated by using the highest 1 (maximum), means of the highest 2, 3, 4, and 5 Schmidt hammer rebound values (SH_1, SH_2, SH_3, SH_4, and SH_5) instead of SH_{MAX} and SH_{10} (mean of the ten Schmidt hammer readings [one set] performed in the field on a certain location) instead of SH_{MEAN} as in Equation 6.2.

$$SHRF_i = \left[\frac{SH_i - SH_{10}}{SH_{10}} \right] \cdot 100, \quad (\%) \tag{6.2}$$

where, indice (i) is 1, 2, 3, 4, and 5 indicating the number of the highest Schmidt hammer values used for estimations or percentage/percentile of the highest Schmidt hammer values in one set of ten readings. For example, $i = 3$ means the highest 30% of the data or the 3rd percentile of the data.

The effect of cleat spacings (frequencies) on $SHRF_i$ values were statistically investigated. Face cleat spacings and Schmidt hammer reduction factors in different percentiles are given in Table 6.1 for nine different coal samples. As seen in Figure 6.4, the best correlation with face cleat spacing is obtained for the mean values of the highest three Schmidt hammer readings. Figure 6.4 enables calculation of the cleat spacings and Schmidt hammer reduction factors in 134 locations, where Schmidt hammer tests were carried out.

TABLE 6.1

Schmidt Hammer Reduction Factors for Different Face Cleat Spacings of the
Coal Samples

Sample ID	FC_S (cm)	$SHRF_1$	$SHRF_2$	$SHRF_3$	$SHRF_4$	$SHRF_5$
21	1.25	23.5	21.4	19.8	18.6	16.7
26	0.9	24.4	22.4	20.8	18.8	16.0
27	0.5	60.9	39.5	27.3	20.8	14.2
30	0.75	27.7	23.4	19.7	16.5	14.4
41a	0.7	33.0	25.9	20.4	17.1	13.5
41b	1.3	55.6	26.2	20.5	17.4	13.8
42	1.5	25.3	18.5	15.5	13.4	10.7
43	2.0	23.0	10.4	9.5	9.5	7.2
45	1.5	16.6	15.8	13.5	9.9	7.8

Source: Bilgin, N. et al., *Int. J. Rock Mech. Min. Sci.,* 84, 25–33, 2016.
FC_S: Face Cleat Spacing.
$SHRF_1$, $SHRF_2$, $SHRF_3$, $SHRF_4$, and $SHRF_5$ are the Schmidt hammer Reduction Factors estimated
by using the highest (maximum), mean values of the highest 2, 3, 4, and 5 Schmidt hammer
readings (SH_1, SH_2, SH_3, SH_4, and SH_5), respectively.

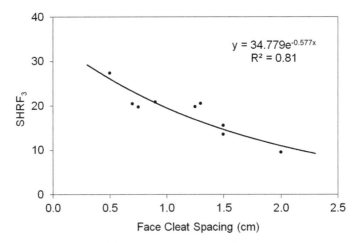

FIGURE 6.4
Relationship between face cleat spacing and Schmidt hammer reduction factor ($SHRF_3$) esti-
mated for mean of the highest three Schmidt hammer readings samples. (From Bilgin, N. et al.,
Int. J. Rock. Mech. Min. Sci., 84, 25–33, 2016.)

6.5 Relationships between Schmidt Hammer Rebound Values and Coal Strength Properties

Sheorey et al. (1984), Haramy and Demarco (1995), and Ghose and Chakraborti (1986) established the following relations: UCS = 0.4RN − 3.6, UCS = 0.994RN − 0.383, UCS = 0.88RN − 12.11, respectively. For coal, these equations have relative correlation coefficients of 0.94, 0.70, and 0.87 respectively. In these equations, RN is Schmidt hammer rebound value (N type) and UCS is uniaxial compressive strength in MPa.

The most widely used strength property for coal is uniaxial compressive strength. Due to the difficulties in taking core samples, cubic samples are preferred in coal strength testing. It is reported in the literature that uniaxial compressive strength value of coal decreases with increasing cube size and after 50 mm of cube size, the values stay almost constant as discussed in Chapter 5.

The 49 representative coal samples were obtained from the locations where Schmidt hammer tests were realized and cube samples having sizes of approximately 50 mm × 50 mm × 50 mm were prepared for uniaxial compressive strength, acoustic wave velocities (P and S), dynamic elasticity modulus and Poisson's ratio, and static elasticity modulus tests. However, the applicability of determining the P and S wave velocities using small cubic specimens should be considered depending on cleat frequency. The size of the specimen should be greater than the frequency of the cleats. Samples were carefully prepared to have smooth and parallel surfaces and the loading direction was kept perpendicular to the bedding planes for all samples. Physical and mechanical properties of the samples are summarized in Table 6.2. In this table, the mean values of ten readings and mean values of the three highest readings of Schmidt hammer rebound values are also given since it is shown that the mean of the highest three Schmidt hammer rebound values within one set of Schmidt hammer tests (ten readings) would give the best correlation with coal strength values.

The relationships between mean Schmidt hammer values (of ten readings at a certain location), mean of the highest three Schmidt hammer values and uniaxial compressive strength, dynamic elasticity modulus, and static elasticity modulus values are given in Figures 6.5 through 6.7. As seen in these figures, rock strength values are best correlated with Schmitt hammer values if the highest three Schmidt hammer rebound values are used. This is one of the main conclusions of this research study suggesting that in coal strength characterization using Schmidt hammer values should be represented with the mean values of the highest three values out of ten readings on a certain location.

TABLE 6.2

Physical and Mechanical Properties of the Coal Samples

Sample ID	ρ (g/cm³)	UCS (MPa)	E_{STA} (GPa)	PW_V (m/s)	SW_V (m/s)	E_{DYN} (GPa)	ν_{DYN}	SH_{10}	SH_3
1	1.76	89.1	4.51	3,587	1,394	9.67	0.41	59.1	66.5
7	1.70	38.5	—	—	—	—	—	45.3	60.1
9	1.70	19.2	—	—	—	—	—	33.9	49.0
10	1.68	8.0	—	—	—	—	—	33.8	40.3
11	1.72	26.1	—	—	—	—	—	55.9	62.0
12	1.70	5.9	—	—	—	—	—	34.2	37.1
13	1.70	14.1	—	—	—	—	—	39.8	42.0
14	1.72	8.2	—	—	—	—	—	36.9	41.2
15	1.72	7.1	—	—	—	—	—	35.1	41.3
16	1.80	19.4	1.14	2,471	968	4.75	0.41	51.0	59.0
18	1.72	8.4	0.50	2,462	793	3.12	0.44	26.9	30.6
20	1.68	55.7	2.46	2,450	1,250	6.94	0.32	42.9	49.3
22	1.66	102.3	—	2,644	1,027	4.93	0.41	62.2	74.0
26	1.70	14.4	1.68	2,707	1,601	10.75	0.23	40.2	48.6
27	1.70	32.0	—	—	—	—	—	46.6	59.3
32	1.70	7.8	0.99	2,580	1,048	5.24	0.40	34.8	41.3
35	1.72	14.0	—	2,659	985	4.74	0.42	31.6	37.7
36	1.72	35.0	2.45	—	—	—	—	46.0	51.6
38	1.72	36.1	2.45	—	—	—	—	44.3	58.5
40	1.32	26.0	2.12	2,753	1,281	5.90	0.36	35.2	49.7
43	1.70	49.6	—	—	—	—	—	44.4	48.7
47	1.70	17.9	—	—	—	—	—	40.2	44.5
48	1.70	24.0	—	—	—	—	—	52.3	60.0

Source: Bilgin, N. et al., *Int. J. Rock Mech. Min. Sci.*, 84, 25–33, 2016.

ρ: Unit weight (natural), UCS: Uniaxial compressive strength, E_{STA}: Static elasticity modulus, E_{DYN}: Dynamic elasticity modulus, PW_V: P wave velocity, SW_V: S wave velocity, ν_{DYN}: Dynamic Poisson's ratio, SH_{10}: Mean of the ten Schmidt hammer readings (one set) measured in the field on a certain location, SH_3: Mean of the highest three values within one set of Schmidt hammer readings measured in the field.

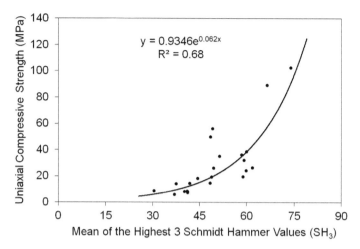

FIGURE 6.5
Relationship between uniaxial compressive strength and mean values of the highest three Schmidt hammer readings. (From Bilgin, N. et al., *Int. J. Rock. Mech. Min. Sci.*, 84, 25–33, 2016.)

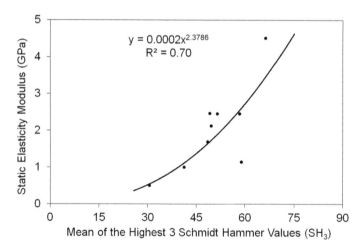

FIGURE 6.6
Relationship between static elasticity modulus and mean values of the highest three Schmidt hammer readings. (From Bilgin, N. et al., *Int. J. Rock. Mech. Min. Sci.*, 84, 25–33, 2016.)

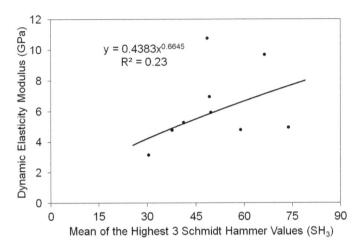

FIGURE 6.7
Relationship between dynamic elasticity modulus and mean values of the highest three Schmidt hammer readings. (From Bilgin, N. et al., *Int. J. Rock. Mech. Min. Sci.*, 84, 25–33, 2016.)

The Schmidt hammer values given in Table 6.1 are grouped according to the location of the Schmidt hammer readings as presented in Table 6.2, which provides a guide for planning mechanized mining in locating longwall faces and making decisions for drill and blast.

Table 6.3 gives a classification system for excavability of the coal based on the mean of the highest three Schmidt hammer rebound values (SH_3), Schmidt hammer reduction factor ($SHRF_3$) estimated based on SH_3, explosive specific charge realized in the studied mine, uniaxial compressive strength, and specific energy values obtained by small-scale linear cutting tests in the laboratory in unrelieved cutting mode with a standard chisel tool at 5 mm depth of cut. Specific energy is the most commonly used criteria to define excavatability of coal, the values given in Table 6.3 are taken from a paper recently published by the authors of this study (Bilgin et al. 2015).

TABLE 6.3

Excavatability Classification of Coal Seams Related to Different Criterion

Excavatability Classification	SH_3	$SHRF_3$	Field Explosive Consumption (kg/m³)	Uniaxial Compressive Strength (MPa)	Field Specific Energy (MJ/m³)
Easy	<48	30–59	<0.35	<18	1.0–1.5
Moderate	48–54	20–30	0.35–0.40	18–27	1.5–2.0
Difficult	54–65	14–20	0.40–0.55	27–90	2.0–3.0

Source: Bilgin, N. et al., *Int. J. Rock Mech. Min. Sci.*, 84, 25–33, 2016.
SH_3: Mean of the highest three values within one set of Schmidt hammer readings, $SHRF_3$: Schmidt hammer reduction factor estimated by using SH_3.

6.6 Conclusive Remarks

The Schmidt hammer test is widely used in the laboratory and in the field to make quick estimations of rock strength and is well correlated with some mechanical rock properties. However, if it is used for the characterization of coal seams, there is no doubt that the tests will be affected by the presence of cleats. Although there are several published research works related to the application of Schmidt hammer on coal seams, the effect of cleat spacing on the test values has not been established till the previous paper published by the authors in Bilgin et al. (2016). This work carried out on the methodology of using the Schmidt hammer reveals the fact that strength reduction factor defined by Young and Fowell (1978) for discontinuities in sedimentary rocks may be applied for also coal seams with certain cleat spacing if the mean values of the highest three Schmidt hammer rebound values within a set of ten readings at a certain location are used. Uniaxial compressive strength and static elasticity modulus of coal are also found to be correlated well if the mean of the highest three Schmidt hammer rebound values are considered instead of using mean Schmidt hammer rebound values of ten readings. By using this methodology, coal seams can be reliably classified in term of excavatability/ workability. It is important also to note that the excavatability classification proposed in this chapter is closely related to the explosive specific charges used in the mine for different horizons of the coal seam indicating a verification for the suggested excavatability classification. However, it should be mentioned that measuring cleat spacing is a difficult task. In this study, the cleats being apparent from visual inspection of the coal samples in the laboratory were considered in the statistical analysis. It is strictly advised to do further research studies to improve the excavatability classification proposed in this chapter.

Based on the studies performed, a classification for excavatability/workability of coal seams has been developed to help designers for the future mine planning activities.

References

ASTM 2015. Standard test method for determination of rock hardness by rebound hammer method. *Designation*: D5873–14, 6 p.

Ayday, C. and Goktan, R. M. 1992. Correlations between L and N-type Schmidt hammer rebound values obtained during field testing. *ISRM Symposium on Rock Characterization, Eurock 92*, Editors Hudson, J. A., Chester, UK, pp. 47–50.

Aydin, A. 2009. ISRM suggested method for determination of the Schmidt hammer rebound hardness-revised version. *International Journal of Rock Mechanics and Mining Sciences*, 6:627–663.

Aydin, A. and Basu, A. 2005. The Schmidt hammer in rock material characterization. *Engineering Geology*, 81(1):1–14.

Basu, A. and Aydin, A. 2004. A method for normalization of Schmidt hammer rebound values. *International Journal of Rock Mechanics and Mining Sciences*, 41(7):1211–1214.

Bilgin, N., Balci, C., Copur, H., Tumac, D., and Avunduk, E. 2015. Cuttability of coal from the Soma coalfield in Turkey. *International Journal of Rock Mechanics and Mining Science*, 73:123–129.

Bilgin, N., Copur, H., and Balci, C. 2016. Use of Schmidt hammer with special reference to strength reduction factor related to cleat presence in a coalmine. *International Journal of Rock Mechanics and Mining Sciences*, 84:25–33.

Bilgin, N., Dincer, T., and Copur, H. 2002. The performance prediction of impact hammer from Schmidt hammer rebound values in Istanbul metro tunnel drivages. *Tunneling and Underground Space Technology*, 17(3):237–247.

Bilim, N. and Ozkan, I. 2008. Determination of the effect of roof pressure on coal hardness and excavation productivity: An example from a Çayirhan lignite mine, Ankara, Central Turkey. *International Journal of Coal Geology*, 75(2):113–118.

Bruno, G., Vessia, G., and Bobbo, L. 2013. Statistical method for assessing the uniaxial compressive strength of carbonate rock by Schmidt hammer tests performed on core samples. *Rock Mechanics and Rock Engineering*, 46:199–206.

Buyuksagis, I. S. and Goktan, R. M. 2007. The effect of Schmidt hammers type on uniaxial compressive strength prediction of rock. *International Journal of Rock Mechanics and Mining Science*, 44(2):299–307.

Carter, P. G. and Sneddon, M. 1977. Comparison of Schmidt hammer, point load and unconfined compression tests in carboniferous strata. *Proceedings on Rock Engineering*, University of Newcastle Upon Tyne, Tyne, UK, pp. 197–210.

Cobanoglu, I. and Celik, S. B. 2008. Estimation of uniaxial compressive strength from point load strength, Schmidt hardness and P-wave velocity. *Bulletin of Engineering Geology and the Environment*, 67(4):491–498.

Dawson, G. K. W. and Esterle, J. S. 2010. Controls on coal cleat spacing. *International Journal of Coal Geology*, 82(3–4):213–328.

Demirdag, S., Yavuz, H., and Altindag, R. 2009. The effect of sample size on Schmidt rebound hardness value of rock. *International Journal of Rock Mechanics and Mining Science*, 46(4):725–730.

Dincer, I., Acar, A., Cobanoglu, I., and Uras, Y. 2004. Correlation between Schmidt hardness, uniaxial compressive strength and Young's modules for andesites, basalts and tuffs. *Bulletin of Engineering Geology and the Environment*, 63(2):141–148.

Evans, I. and Pomeroy, C. D. 1966. *The Strength, Fracture and Workability of Coal*. Pergamon Press, London, UK.

Fener, M., Kahraman, S., Bilgil, A., and Gunaydin, O. 2005. A comparative evaluation of indirect methods to estimate the compressive strength of rocks. *Rock Mechanics and Rock Engineering*, 38(4):329–333.

Fowell, R. J. and McFeat-Smith, I. 1976. Factors influencing the cutting performance of a selective tunnelling machine. *Proceedings of the International Symposium IMM (Tunelling'76)*, IMM, London, UK, pp. 301–309.

Ghose, A. K. and Chakraborti, S. 1986. Empirical strength indicates of Indian coals-an investigation. *Proceeding of 27th US Symposium on Rock Mechanics*. Editor. Hartman, H. L. University of Alabama, Balkema, Rotterdam, the Netherlands, pp. 59–61.

Goktan, R. M. and Ayday, C. 1993. A suggested improvement to the Schmidt rebound hardness ISRM suggested method with particular reference to rock machineability. *International Journal of Rock Mechanics and Mineral Science and Geomechanics Abstracts*, 30(3):321–322.

Greco, R. and Sorriso-Valvo, M. 2005. Relationships between joint apparent separation, Schmidt hammer rebound value, and distance to faults, in rocky outcrops, Calabria, Southern Italy. *Engineering Geology*, 78(3–4):309–320.

Haramy, K. Y. and DeMarco, M. J. 1995. Use of Schmidt hammer for rock and coal testing. *26th US Symposium on Rock Mechanics*. Editor. Ashworth, E., CRC Press, Boca Raton, FL. ISBN 9789061916017, pp. 549–555.

Kahraman, S. 2001a. A correlation between P-wave velocity, number of joints and Schmidt hammer rebound number. *International Journal of Rock Mechanics and Mining Science*, 38(5):729–733.

Kahraman, S. 2001b. Evaluation of simple methods for assessing the uniaxial compressive strength of rock. *International Journal of Rock Mechanics and Mining Science*, 38(7):981–994.

Kahraman, S., Fener, M., and Gunaydin, O. 2002. Prediction the Schmidt hammer values of in-situ intact rock from core sample values. *International Journal of Rock Mechanics and Mining Science*, 39(3):395–399.

Karaman, K. and Kesimal, A. 2015. A comparative study of Schmidt hammer test methods for estimating the uniaxial compressive strength of rocks. *Bulletin of Engineering Geology and the Environment*, 74(2):507–520.

Katz, O., Reches, Z., and Roegiers, J. C. 2000. Evaluation of mechanical rock properties using a Schmidt hammer. *International Journal of Rock Mechanics and Mining Science*, 37:723–728.

Kidybinski, A. 1968. Rebound number and the quality of mine roof strata. *International Journal of Rock Mechanics and Mineral Science and Geomechanics Abstracts*, 5(4):283–292.

Laubacha, S. E., Marrettb, R. A., Olson, J. E., and Scotta, A. R. 1998. Characteristics and origins of coal cleat: A review. *International Journal of Coal Geology*, 35(1–4):175–207.

McCulloch, M. C., Deul, M., and Jeran, P. W. 1974. Cleat in bituminous coalbeds. USA Bureau of Mines Report of Investigations 7910, Pittsburgh Mining and Safety Research Center, Pittsburgh, PA, 24 p.

Minaeian, B. and Ahangari, K. 2013. Estimation of uniaxial compressive strength based on P-wave and Schmidt hammer rebound using statistical method. *Arabian Journal of Geosciences*, 6(6):1925–1931.

Niedzielski, T., Migon, P., and Placek, A. 2009. A minimum sample size required from Schmidt hammers measurement. *Earth Surface Processes and Landforms*, 34(13):1713–1725.

Ozbek, A. 2009. Variation of Schmidt hammer values with imbrication direction in clastic sedimentary rocks. *International Journal of Rock Mechanics and Mining Science*, 46(3):548–554.

Ozkan, I. and Bilim, N. 2008. A new approach for applying the in-situ Schmidt hammer test on a coal face. *International Journal of Rock Mechanics and Mining Science*, 45(6):888–889.

Poole, R. W. and Farmer, I. W. 1978. Geotechnical factors acting on tunnelling machine performance in coal measures rocks. *Tunnel Tunnelling*, 15:27–30.

Poole, R. W. and Farmer, I. W. 1980. Consistency and repeatability of Schmidt hammer rebound data during field testing. *International Journal of Rock Mechanics and Mineral Science and Geomechanics Abstracts*, 17(3):167–171.

Roxborough, F. F. and Whittaker, B. N. 1964. Roof control and coal hardness. *Colliery Engineering*, December, 511–517.

Roxborough, F. F. and Whittaker, B. N. 1965. Roof control and coal hardness. *Colliery Engineering*, January, 19–24.

Sachpazis, C. I. 1990. Correlating Schmidt hammer hardness with compressive strength and Young's modulus of carbonate rocks. *Bulletin of the International Association of Engineering Geology*, 42(1):75–83.

Saptono, S., Kramadibrata, S., and Sulistianto, B. 2013. Using the Schmidt hammer on rock mass characteristic in sedimentary rock at Tutupan coal mine. *Procedia Earth and Planetary Science*, 6:390–395.

Sengun, N., Altindag, R., Demirdag, S., and Yavuz, H. 2011. P-wave velocity and Schmidt rebound hardness value of rocks under uniaxial compressional loading. *International Journal of Rock Mechanics and Mining Sciences*, 48(3):693–696.

Sharma, P. K., Kandelwal, M., and Singh, T. N. 2011. A correlation between Schmidt hammer rebound numbers with impact strength index, slake durability index and P-wave velocity. *International Journal of Earth Sciences*, 100(1):189–195.

Shorey, P. R., Barat, D., Das, M. N., Mukherjee, K. P., and Singh, B. 1984. Schmidt hammer rebound data for estimation of large scale in situ coal strength. *International Journal of Rock Mechanics and Mineral Science and Geomechanics Abstracts*, 21(1):39–42.

Tandon, R. S. and Gupta, V. 2015. Estimation of strength characteristics of different Himalayan rocks from Schmidt hammer rebound, point load index, and compressional wave velocity. *Bulletin of Engineering Geology and the Environment*, 74(2):521–533.

Tumac, D. 2015. Predicting the performance of large diameter circular saws based on Schmidt hammer and other properties for some Turkish carbonate rocks. *International Journal of Rock Mechanics and Mining Sciences*, 75:159–168.

Ulusay, R. 2015. *The ISRM Suggested Methods for Rock Characterization, Testing and Monitoring (2007–2014)*. Springer, Berlin, Germany. ISBN 978-3-319-07712-3, 293 p.

Vakili, A. and Hebblewhite, B. K. 2010. A new cavability assessment criterion for longwall top coal caving. *International Journal of Rock Mechanics and Mining Science*, 47(8):1317–1329.

Yagiz, S. 2009. Predicting uniaxial compressive strength, modulus of elasticity and index properties of rocks using the Schmidt hammer. *Bulletin of Engineering Geology and the Environment*, 68(1):55–63.

Yasar, E. and Erdogan, Y. 2004. Estimation of rock physico mechanical properties using hardness methods. *Engineering Geology*, 71(3–4):281–288.

Yilmaz, I. and Sendir, H. 2002. Correlation of Schmidt hardness with unconfined compressive strength and Young's modulus in gypsum from Sivas (Turkey). *Engineering Geology*, 66(3–4):211–219.

Young, R. P. and Fowell, R. J. 1978. Assessing rock discontinuities. *Tunnels and Tunnelling*, 10(5):45–48.

Wang, H., Lin, H., and Cao, P. 2017. Correlation of UCS rating with Schmidt hammer surface hardness for rock mass classification. *Rock Mechanics and Rock Engineering*, 50:195–203.

7

Cutting of Coal, Coal Cutting Mechanics, Laboratory Coal Cutting Experiments

7.1 Introduction

Coal has been mined in various parts of the world throughout history and coal mining continues to be an important economic activity today. Compared to wood fuels, coal yielded a higher amount of energy per mass and could be obtained in areas where wood was not readily available. Though historically used as a means of household heating, coal is now mostly used in industry, especially in the area of metallurgy, as well as electricity generation.

Mechanization is an important strategy in the design and operation of modern mines. The objectives of mine mechanization include improved safety conditions, higher productivity, and a reduction in mining costs. The successful application of mechanization typically results in fewer people being employed directly in the mining. Longwall-mining machinery increased in popularity since 1960s, producing several times more coal with less workers, compared to hand mining methods. Eliminating the need for room and pillar mining method, plows and shearers cut the coal from the face of an entire panel into an automatically advancing roof support system.

In this chapter, to make the subject clearer, it is aimed to summarize the historical development of coal cutting machines, the cutters used, the cutting theories, and the laboratory coal cutting experiments to understand better the coal cutting mechanics toward designing more efficient modern coal getting machines.

7.2 Historical Development of Coal Cutting Machines

For a more detailed information on this topic, the readers of this book are advised to read the monumental work written by Stack (1995) entitled "Encyclopedia of Tunnelling, Mining and Drilling Equipment." However, a brief summary of historical development of coal cutting machines will also be given below.

The Industrial Revolution, which began in Britain in the eighteenth century, and later spread to continental Europe, North America, and Japan, was based on the availability of coal to power steam engines. International trade expanded dramatically when coal-fed steam engines were built for the railways. Coal was cheaper and much more efficient than wood fuel in most steam engines. As central and northern England contained an abundance of coal, many mines were situated in these areas, as well as the South Wales coalfield and Scotland. The small-scale techniques were unsuited to the increasing demand, with extraction moving away from surface extraction to deep shaft mining as the Industrial Revolution progressed.

The mechanization of coal mining with cutting technology occurred as early as 1860. The first cutting machines were simple devices comprised of circular saws with picks positioned around the machine. It took nearly an entire century, however, for coal cutting equipment to be developed to a level sophisticated enough that miners could abandon hand tools.

By 1900, different types of undercutting machines eliminated the time-consuming task of undercutting by picks. However, miners still continued to drill, blast, and load manually, and control of these critical skills enabled them to maintain much of their independence at the working face.

One of the early chain saw cutters is seen in Figure 7.1, and one latest type of these machines which was manufactured by Anderson Boyes and exhibited in the Snibston Discovery Museum is seen in Figure 7.2.

Introduction of the armored face conveyor into the United Kingdom mines in the late 1940s enabled continuous mining. About 1945, the company Anderson Boyes became the first manufacturer of effective longwall cutter loaders, which were the first machines capable both of cutting the coal and loading it onto the belt conveyor.

In the early 1950s, Anderson Boyes developed the trepanner to meet the National coal board (NCB's) need for a power loader capable of producing the large size coal required for boilers, particularly for railways, and have dropped out of use as this need disappeared. A typical trepanner is seen in Figure 7.3.

One of the first coal shearing machines was manufactured in 1960 by Anderson Boyes, the company also provided the basic concept from which the NCB developed and patented the principle of the Anderson shearer loader, which was the forerunner of the modern shearer. The company changed its name to Anderson Strathclyde Ltd. thereafter.

FIGURE 7.1
Miner using a coal cutting machine to undercut the coal seam (www.ncm.org.uk from IMH Group, https://www.ncm.org.uk/downloads/195/Mine_machines_-_coal_cutting.pdf).

FIGURE 7.2
Early Anderson Boyes coal cutter exhibited at the Snibston Discovery Museum, https://www.gracesguide.co.uk/Anderson,_Boyes_and_Co.

Germany is also one of the countries which was influenced tremendously from Industrial Revolution and coal production increased steadily. In 1914, Eickhoff in Germany, founded by Johann Heinrich Carl Eickoff, built the first bar cutting machine in continental Europe. The first hydraulic cutting machine was launched in 1950, and the first longwall shearer

FIGURE 7.3
A trepanner cutting along a coal seam (www.ncm.org.uk from IMH Group https://www.ncm.org.uk/downloads/195/Mine_machines_-_coal_cutting.pdf).

loader was introduced in 1954. The mining machinery division of the company is the largest division with around 300 shearers running currently worldwide. Eickhoff has also expanded into the traditional room and pillar mining market with its own continuous miner in South Africa, Belorussia, China, and Russia (https://www.scrapmonster.com/company/eickhoff-corporation/31681).

Plowing is a coal mining method invented in the early 1940s. Konrad Grebe was the inventor of the first built plow, installed in Germany in Ibbenbüren Mine in 1941. The plow was then redesigned by Wilhelm Löbbe, chief engineer at Westfalia Lunen (a predecessor company of Caterpillar) back in 1947 in an attempt to modernize and mechanize underground coal mining. Löbbe also invented the PANZER armored face conveyor, which was successfully installed underground for the first time in Poland in 1942. Löbbe improved plow performance by reducing the cutting depth and increasing the cutting speed. The "fast plow" was first installed in 1949 at Friedrich-Heinrich Mine in Germany. With an installed power of only 2 × 40 kW, it achieved a daily output of more than 1,000 tons for the first time in 1950. Many different plow models were designed and tested underground, until in the mid-1990s by Westfalia

Lunen. Horsepower and plow speed have increased steadily over time, but it was not until 1989 that the biggest weakness of plow systems was overcome: automated plowing using electro-hydraulic controls with defined cutting depths finally allowed plowing to become one of the most productive mining method for seams below 1.8 m thick. This technology step, like most inventions in plowing, came from Caterpillar predecessor Westfalia Lunen (Paschedag 2014).

In the United States of America, the first coal cutting machine was designed by Jonas P. Mitchell, who applied for a patent in 1891. By 1893, Mitchell built his forth machine. A few years later, the Jeffrey Manufacturing Company built also similar type machines.

In 1950, the Eastern Associated Coal Corporation of West Virginia began using a coal plow for longwall mining, from that time the number of longwalls gradually expended. To meet this demand, Joy Manufacturing Company began selling longwall cutters of various types toward the end of 1960s. Thereafter, the company designed and manufactured new generations of shearers for longwalls and continuous miners for room and pillar mining (Stack 1995).

New generations of plows, shearers, and continuous miners used in coal mining are shown in Figures 7.4 through 7.6.

FIGURE 7.4
A plow used in a longwall. (Courtesy of Caterpillar.)

FIGURE 7.5
A shearer, SL300 manufactured by Eickhoff, used in a coal mine. (Courtesy of Eickhoff.)

FIGURE 7.6
A continuous miner used in room and pillar mining method. (Courtesy of Sandvik.)

7.3 Coal Cutting Tools, Chisel Tools/Radial Cutters, Point Attack Tools/Conical Cutters

The cutting tools used in coal getting machines can be classified as chisel cutters/radial cutters, point attack tools/conical cutters. Typical chisel/radial cutters and point attack/conical cutters are illustrated in Figures 7.7 and 7.8. Conical cutter and its installation in a shearer drum and in a roadheader cutterhead are given in Figure 7.9.

Many parameters should be considered when choosing a tool for a ground type, such as type, specifications, capacity of the machine; cuttability, strength, abrasiveness, texture, hard mineral content of the ground; tool lacing on cutterhead; tool consumption rate; and cutting capability/performance of the cutting tool. The selection of a suitable cutting tool, which also leads to an optimized design of the cutterhead and mechanical miner, is mainly based on comparison of their cutting performance and tool costs for a given ground/formation. Tool consumption, which is mainly related to the abrasivity of the coal, is one of the main factors dictating the cutting efficiency of the machine, since checking and replacing the cutters effect in most cases the machine utilization time. A typical example is given for Kuzbass coalfield in Russia. A coal mine in average spends 15,000 cutters–18,000 cutters per year which increases the cost of mining and a tremendous effort is spend to prolong the cutter lives (Prokopenko et al. 2015).

A recent research study on the wear of point attack tools revealed the following important points (Liu et al. 2017): The larger the height of the carbide tip and cone angle of pick tip were, the easier it was to protect the head

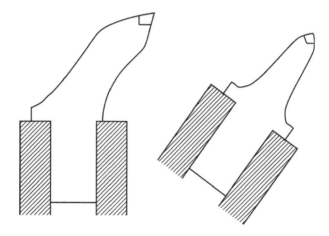

FIGURE 7.7
Radial and forward attack tools. (From Bilgin, N. et al., *Mechanical Excavation in Mining and Civil Industries*, Boca Raton, FL, CRC Press/Taylor & Francis Group, 2014.)

FIGURE 7.8
Conical cutters with different geometrical characteristics. (Courtesy of Sandvik.)

FIGURE 7.9
Conical cutter and its installation in a shearer and in a roadheader. (From Liu, S. et al., *Eng. Fail. Anal.*, 74, 172–187, 2017.)

face of the pick body from interference with coal-rock and avoid carbide tip loss due to wear of the pick body. The suitable height of the carbide tip was found to be 20 mm–24 mm, and the cone angle of approximately 80°. Pick wear decreased with cutting (attack) angle, but the cutting angle had to be 45°–50°, when considering pick cutting force. Cutter wear increased with increasing rake angle. It was also found that the mean peak pick cutting torque of a shearer increased with wear degree and the largest increase percentage was 30%.

It is a fact that the severe abrasive wear of the current cemented tungsten carbide tools is a "bottleneck" that limits the usage of machinery in abrasive rock and coal. It is important to note here that to address this issue, a revolutionary thermally stable diamond composite based cutting tool, also called super material abrasive resistant tool was developed by Commonwealth Scientific and Industrial Research Organisation, and ready for industrial applications (Shao et al. 2017).

A general classification of drag tools based on the type of cutting action and application limits in terms of abrasivity and strength of rock masses is given in Table 7.1 (Copur et al. 2012). The cutting action of drag tools, which limits their utilization up to moderately abrasive medium-strength rocks (up to 100 MPa–120 MPa of uniaxial compressive strength), is basically dragging (scratching) with a high friction over the excavated surface, making them prone to wearing out easily. A drag tool keeps its durability mostly up to 4 tons–5 tons of loads. Drag tools have a rectangular shank to be fixed on the tool holder, except for conical tools, which have cylindrical shanks that provide an even wearing of tool tips, giving a longer tool life compared to the other types of drag tools.

Radial tools are normally used with continuous miners, shearer-loaders, and roadheaders for the excavation of non-abrasive soft grounds and minerals such as coal, salt, and potash (up to 40 MPa–60 MPa of uniaxial

TABLE 7.1

General Classification of Drag Tools Based on Their Action Types

Action	Subaction	Cutting Tools	Operational Limits, UCS[a] and Abrasivity
Dragging	Constant, fixed	Radial	<40–60 MPa, non-abrasive rocks
		Scraper	<60–80 MPa, non-abrasive rocks
		Chisel	<80–100 MPa, low abrasive rocks
	Rotating	Conical	<100–120 MPa, moderately abrasive rocks

Source: Copur, H. et al., Predicting cutting performance of chisel tools by using physical and mechanical properties of natural stones, *Proceedings of the European Rock Mechanics Symposium (Eurock 2012, ISRM International Symposium)*, May 28–30, Stockholm, Sweden, 14 p, 2012.

[a] Uniaxial compressive strength

compressive strength). Their bodies are made of hardened steel, and their insert tips are made of tungsten carbide, providing for longer tool life against wearing. Although they are the most efficient tools, their tool shapes make them more prone to wearing and having lower operational life than that of conical cutters.

7.4 Cutting Theories

A researcher or a mining engineer dealing with mechanical excavation of coal should first understand the basic principles of coal cutting mechanics affecting the coal cutting process. The efficiency of coal cutting machines depends on choosing right cutting tools with correct geometrical and operational parameters, such as depth of cut, the design of tool arrays, and their correct position in the cutting head. The inefficient torque and thrust may cause high specific energy, high amount of dust, etc. On the other hand, high forces acting on a drag tool may result in gross fracture damage to the tungsten carbide cutting tip or tool material and also damage to the machine components by exceeding the machine's torque and thrust capacities. Therefore, it is essential to understand the basic aspects of coal cutting mechanics to minimize the large cost of a trial-and-error approach with an excavation machine in the field. The coal/rock-breakage mechanism is complex in nature. Experimental and theoretical studies are still continuing for a better understanding of rock/coal cutting mechanics. The section given below is a summary of a chapter, of which has already been published by the authors (Bilgin et al. 2014). The readers are also advised to read a recent publication to understand the subject better (Copur et al. 2017).

7.4.1 Simple Chisel Tools

The independent and dependent variables when cutting with chisel cutters are defined in Figure 7.10.

Basic independent variables are as follows:

d = Depth of cut;
w = Width of tool;
α = Rake angle;
β = Clearance angle; and
s = Cutter spacing.

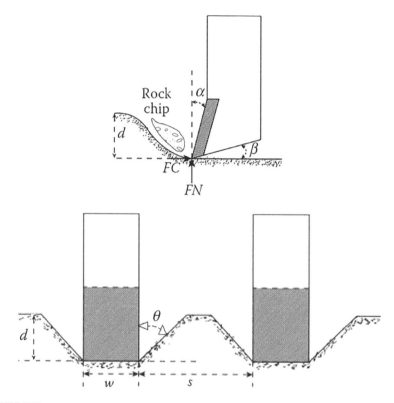

FIGURE 7.10
Dependent and independent variables when cutting with chisel cutters. (Modified after Bilgin, N. et al., *Mechanical Excavation in Mining and Civil Industries*, Boca Raton, FL, CRC Press/Taylor & Francis Group, 2014.)

Basic dependent variables are as follows:

FC = Cutting force (force acting in the direction of cutting action);

FN = Normal force (force acting perpendicular to the direction of cutting action);

θ = Breakout angle;

Q = Yield (the volume of material cut during a unit length of cut); and

SE = Specific energy which is defined as the energy spent to cut a unit volume of rock, it is found dividing FC by yield.

The force acting on a cutting tool changes constantly in magnitude during the cutting process due to chipping and brittle nature of the coal/rock. Averages of all of the force changes during the course of the cutting action give the mean cutter force, mean peak forces are averages of the peak forces for a given cutting condition. The ratio of peak forces to mean forces usually

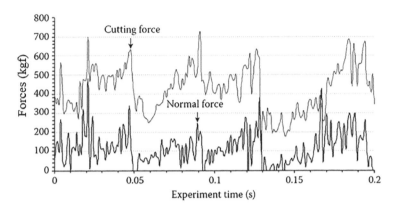

FIGURE 7.11

Typical recorded forces for chisel cutters when cutting. (From Bilgin, N. et al., *Mechanical Excavation in Mining and Civil Industries*, Boca Raton, FL, CRC Press/Taylor & Francis Group, 2014.)

ranges between 1.5 and 3.5, generally being higher with coals. The ratio of cutting force to normal force is around 1–2 for sharp conditions of the tools. However, normal forces are affected by wear more than cutting forces and increase rapidly with tool wear. Typically recorded cutter forces for chisel cutters in sharp condition are seen in Figure 7.11.

The works on coal-cutting mechanics performed by Evans (1962, 1972, 1982, 1984a, 1984b) and Evans and Pomeroy (1966) were used to establish the basic principles of coal cutting, and these have been widely used in the efficient design of excavation machines such as shearers, continuous miners, and roadheaders. Evans, in his studies, demonstrated theoretically that tensile strength was the dominant rock property in rock cutting with chisel tools/picks as formulated in Equation 7.1: where d, w, and α are the parameters as defined in Figure 7.10, σ_t is the tensile strength, and F'C is the peak cutting force.

He also formulated optimum spacing for chisel picks as three to four times the pick width. Roxborough (1973, 1985), Roxborough and Rispin (1973a, 1973b), and Bilgin (1977) suggested that, to some extent, the experimental forces for chisel picks were in good agreement with theoretical values obtained from Equation 7.1.

$$F'C = \frac{2 \cdot \sigma_t \cdot d \cdot w \cdot \sin\frac{1}{2}\left(\frac{\pi}{2} - \alpha\right)}{1 - \sin\frac{1}{2}\left(\frac{\pi}{2} - \alpha\right)} \tag{7.1}$$

7.4.2 Complex Shape Chisel Tools

Radial, forward attack, and point attack tools are illustrated in Figures 7.7 and 7.8. Theoretical works were developed with many simplifications and assumptions and usually for simple chisel cutters and unrelieved cutting mode. Therefore, the theoretical models must be modified for different tool geometries and cutting conditions used in practice, including wear flat, front ridge angle, and v-bottom angle as defined in Figure 7.12.

As explained above, all theoretical works consider the unrelieved cutting mode, taking into account the cutters in isolated (unrelieved, non-interactive, single) mode without any interaction between grooves generated by another tool. However, in a rotary cutterhead of any mechanical miner, the cutting tools are placed in an array where there is always interaction between the groves, generating a relieved cutting condition. There is always an optimum ratio of cutter spacing to depth of cut (s/d) at which the specific energy is minimum. In that case, the energy spent to cut a unit volume of rock is minimum

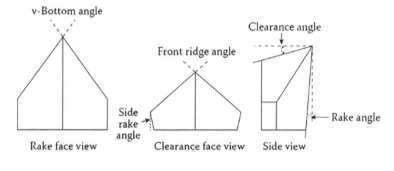

FIGURE 7.12
Basic parameters of complex-shaped chisel cutters.

with minimum cutter consumptions (Roxborough 1973, Roxborough and Rispin 1973a, 1973b). This also makes the tool forces approximately 10% lower than the tool forces obtained in unrelieved cutting mode at the same depth of cuts. Optimum (s/d) ratio is around two for chisel tools and conical cutters when cutting medium-strength rocks. The concept of relieved and unrelieved cutting with chisel cutters is illustrated in Figure 7.13.

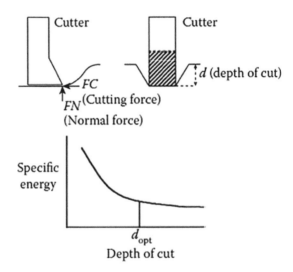

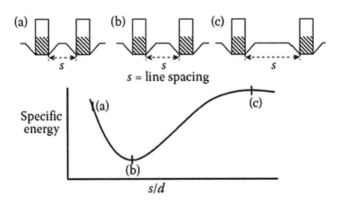

FIGURE 7.13
Basic concept of relieved and unrelieved cutting and relevant parameters. (From Bilgin, N. et al., *Mechanical Excavation in Mining and Civil Industries,* Boca Raton, FL, CRC Press/Taylor & Francis Group, 2014.)

It should also be noted that all theoretical works consider occurrence of a single/major chip, which is the largest/biggest one in a cut. Therefore, all the theoretical models, in fact, estimate the peak/maximum cutting force.

Evans' cutting model, given in Equation 7.1 for chisel tools, was suggested to be modified by Bilgin et al. (2012, 2014) for complex-shaped cutters by using some experimental coefficients as in Equation 7.2:

$$F'C_W = k_1 \cdot k_2 \cdot k_3 \cdot k_4 \cdot \left[\frac{2 \cdot \sigma_t \cdot d \cdot w \cdot \sin \frac{1}{2}\left(\frac{\pi}{2} - \alpha\right)}{2 - \sin \frac{1}{2}\left(\frac{\pi}{2} - \alpha\right)} \right] \tag{7.2}$$

In Equation 7.2, $F'C_W$ is the peak cutting force for a worn, complex-shaped chisel tool, where $k_1 = (F'C_W/F'C_S)$ is a coefficient used for taking into account the effect of wear flat on tool force as defined in Table 7.2, k_2 is a coefficient used for taking into account the effect of front ridge angle, k_3 is a coefficient used for taking into account the effect of v-bottom angle as defined in Table 7.3, k_4 is a coefficient used for taking into account the effect of cutting in relieved mode (usually 0.9), and the other parameters are the same as for Equation 7.1.

TABLE 7.2

Summary of the Effect or Wear Flat on Chisel Tool Forces

Wear Flat (mm)	$F'C_W / F'C_s = k_1$	$F'N_W / F'N_S$	$F'C_W/FC_W$	$F'N_W/FN_W$
0.5	1.27	1.74	2.20	1.74
1.0	1.55	2.41	2.13	1.67
1.5	1.83	3.09	2.06	1.61
2.0	2.11	3.76	1.99	1.54
2.5	2.39	4.43	1.92	1.48
3.0	2.67	5.10	1.85	1.41
3.5	2.95	5.78	1.78	1.35

Note: $F'C_S$, $F'C_W$, $F'N_S$ and $F'N_W$ values are the peak cutting and normal forces for sharp (indices s) and worn (indices w) state of the cutters, respectively; FC_W and FN_W values are the mean cutting and normal forces for worn (indices w) state of the cutters, respectively.

TABLE 7.3

Force Reduction Factors for Front Ridge and v-Bottom Angles to be Used in Modification of Evans' Cutting Theory

Front ridge angle	—	90°	120°	150°	180°
Force reduction factor (k_2)	—	0.65	0.80	0.95	1.00
v-Bottom angle	60°	90°	120°	150°	180°
Force reduction factor (k_3)	0.50	0.65	0.80	0.90	1.00

7.4.3 Point Attack Tools, Conical Cutters

Evans (1984a, 1984b) demonstrated theoretically that tensile strength and uniaxial compressive strength were dominant coal properties in point attack tools as formulated in Equation 7.3.

$$F'C = \frac{16 \cdot \pi \cdot d^2 \cdot \sigma_t^2}{\cos^2(\varnothing/2) \cdot \sigma_c} \tag{7.3}$$

where F'C is the peak cutting force, d is the depth of cut, σ_t is the tensile strength, σ_c is the uniaxial compressive strength and ϕ is the tip angle of the conical cutter.

Goktan (1990) suggested a modification on Evans' cutting theory for point attack tools as indicated in Equation 7.4 and concluded that the force values obtained with this equation were close to previously published experimental values and could be of practical value, if confirmed by additional studies.

$$F'C = \frac{4\pi \cdot \sigma_t \cdot d^2 \cdot \sin^2(\varnothing/2 + \psi)}{\cos(\varnothing/2 + \psi)} \tag{7.4}$$

where ψ is the friction angle between cutting tool and rock and other parameters are as defined above for Equation 7.3.

Roxborough and Liu (1995) also suggested a modification of Evans' cutting theory for point attack tools as given in Equation 7.5. With all the parameters described above, they concluded that for Grindleford sandstone, the predicted mean peak cutting force values are in good agreement with the modified cutting theory. However, the friction angle used was 16° obtained experimentally from a steel block and a natural flat rock surface.

$$F'C = \frac{16\pi \cdot \sigma_c \cdot d^2 \cdot \sigma_t^2}{\left[2\sigma_t + (\sigma_c \cdot \cos(\varnothing/2)) \cdot \left(\frac{1 + \tan\psi}{\tan\left(\psi/2\right)}\right)\right]^2} \tag{7.5}$$

7.5 Cutting Coal with Conical and Radial Cutters

Coal cutting experiments realized by Roepke and Voltz (1983) in Bureau of Mines will serve a basis in this section to explain the basic cutting mechanism of conical and radial cutters and to compare the measured cutting force values with those predicted from Evans' cutting theories. In these comparisons made in this section (as well as in the next section), the estimated peak cutting force from the theory is divided by two to find the mean cutting force, since the cutting theories estimate peak (maximum) cutting force.

The main objective of the research carried out by Roepke and Voltz (1983) was to see the dust generated by different cutters. They tested two conical cutters of 60° tip angles and 90° tip angles and four radial gauge cutters, namely, RAD-1, RAD-2, RAD-3, and RAD-4 which were typical cutters used in continuous miners. RAD-1 and RAD-2 cutters are the most common cutters used in some coal cutting machines. The mean cutter force values generated by RAD-1 and RAD-2 with tool width of 20 mm and rake angle of 10° will be used below for comparative purposes. The coal samples tested were the standard specimen used in Bureau of Mines coal cutting experiments and uniaxial compressive strength of 200 kg/cm² and tensile strength of 13.3 kg/cm² were used in calculating the cutting force from Evans' cutting theory. These values are in good agreement with the values given by Mathey (2015).

The variation of mean cutting force with depth of cut when cutting coal with 60° and 90° conical cutters and radial cutters are given in Figure 7.14. As seen from this figure, the mean cutting force generated by 60° conical cutters and radial cutters are almost same for the depth of cut intervals between 12 mm and 25 mm, which is the interval where the most coal cutting machines work.

The variation of specific energy in unrelieved cutting mode with depth of cut when cutting coal with 60° and 90° conical cutters and radial cutters is seen in Figure 7.15. As expected, specific energy decreases dramaticially with depth of cut and levels off after 12 mm depth of cut. It is interesting to note

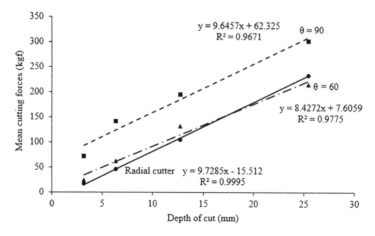

FIGURE 7.14
Variation of mean cutting force with depth of cut in unrelieved cutting mode when cutting coal with 60° and 90° conical cutters and radial cutters. (Adapted from Roepke, W.W. and Voltz, J.I., *Coal-Cutting Forces and Primary Dust Generation Using Tadial Gage Cutters*, Report of Investigations: USA Bureau of Mines, Avondale, MD, 1983.)

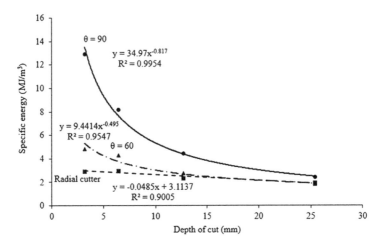

FIGURE 7.15
Variation of specific energy in unrelieved cutting mode with depth of cut when cutting coal with 60° and 90° conical cutters and radial cutters. (Adapted from Roepke, W.W. and Voltz, J.I., *Coal-Cutting Forces and Primary Dust Generation Using Tadial Gage Cutters*, Report of Investigations: USA Bureau of Mines, Avondale, MD, 1983.)

that radial cutters and 60° tip angled conical cutter have almost the same specific energy value after this critical value of depth of cut.

The comparison of measured coal cutting forces with predicted cutting forces from Evans' cutting theory for 60° and 90° conical cutters are given in Figures 7.16 and 7.17. As seen from these figures, for 60°, the measured mean

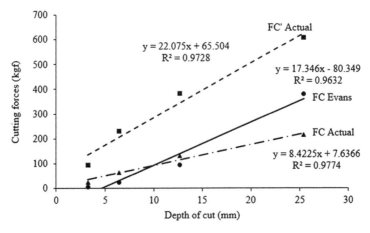

FIGURE 7.16
Comparison of measured coal cutting forces in unrelieved cutting mode with predicted cutting forces from Evans' cutting theory for 60° conical cutter. (Adapted from Roepke, W.W. and Voltz, J.I., *Coal-Cutting Forces and Primary Dust Generation Using Tadial Gage Cutters*, Report of Investigations: USA Bureau of Mines, Avondale, MD, 1983.)

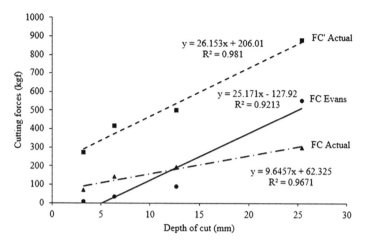

FIGURE 7.17
Comparison of measured coal cutting forces in unrelieved cutting mode with predicted cutting forces from Evans' cutting theory for 90° conical cutter. (Adapted from Roepke, W.W. and Voltz, J.I., *Coal-Cutting Forces and Primary Dust Generation Using Tadial Gage Cutters,* Report of Investigations: USA Bureau of Mines, Avondale, MD, 1983.)

cutting force values are in good agreement with predicted values within practical range of depth of cut which is 12 mm–15 mm depth of cut for most coal cutting machines.

The comparison of measured coal cutting forces with predicted cutting forces from Evans' cutting theory for radial cutters are given in Figure 7.18.

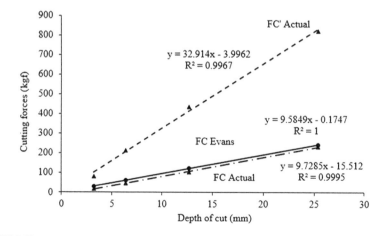

FIGURE 7.18
Comparison of measured coal cutting forces in unrelieved cutting mode with predicted cutting forces from Evans' cutting theory for radial cutters. (Adapted from Roepke, W.W. and Voltz, J.I., *Coal-Cutting Forces and Primary Dust Generation Using Tadial Gage Cutters,* Report of Investigations: USA Bureau of Mines, Avondale, MD, 1983.)

It is not surprising to see in that case the measured mean cutting forces are in good agreement for almost all depths of cut changing from 3 mm to 25 mm.

Cutting force determines the need of torque and power of a cutting machine. Normal force plays an important role in sumping of the cutting head in the coal seam. There is not any theory to predict the normal force when cutting the coal or rock with conical or radial cutters. This is predicted from cutting force/normal force ratios. The coal cutting experiments carried out by Roepke and Voltz (1983) showed that cutting force/normal force ratio went linearly from 1 mm or 3 mm depth of cut up to 2 mm for 25 mm depth of cut for 60° conical cutter. However, this ratio was less in 90° conical cutter changing from 0.7 to 1.2. With radial cutters, cutting force/normal force ratio changed from 1.3 to 4.9 from shallow depth of 3 mm up to deeper depth of 25 mm.

7.6 Cutting Coal with Simple Chisel Tools

The laboratory experiments carried out in Istanbul Technical University, Mining Engineering Department for the coal specimens obtained from the following Turkish coalfields were summarized below in respect to explain the cutting mechanism of simple chisel cutters.

1. Lignite from Milten coalfield (Eskikaya et al. 1995). The compressive strength of the coal was 356 kg/cm^2 and tensile strength was 32.4 kg/cm^2;
2. Hard coal from Amasra A coalfield (Bilgin et al. 2010). The compressive strength of the coal was 141 kg/cm^2 and tensile strength was 0.57 kg/cm^2;
3. Lignite from Soma coalfield (Bilgin et al. 2011, 2015). The compressive strength of the coal specimen was 180 kg/cm^2; and
4. Hard coal from Amasra coalfield B (Balci et al. 2013).

Small scale linear cutting machine was used with a tool width of 12.5 mm, rake angle of −5° and clearance angle of 5°. The tests were performed in both relieved and unrelieved cutting modes.

The variation of cutting force with depth of cut when cutting coal specimen from Amasra A coalfield and Milten coalfield and comparison with the predicted values from Evans' theory are shown in Figures 7.19 and 7.20,

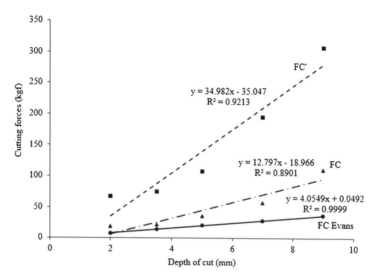

FIGURE 7.19
Variation of cutting force in unrelieved cutting mode with depth of cut when cutting coal specimen from Amasra A coalfield and comparison with predicted values from Evans' theory.

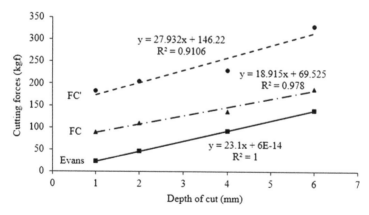

FIGURE 7.20
Variation of cutting force in unrelieved cutting mode with depth of cut when cutting lignite specimen of Milten lignite mine and comparison with predicted values from Evans' cutting theory.

respectively. As seen from these figures, the measured mean cutting forces are almost twice as much as the predicted ones by Evans' theory.

The variation of specific energy in unrelieved cutting mode for different coal specimens shows decreasing trend with depth of cut as expected (Figure 7.21). For relieved cutting mode, specific energy shows an optimum

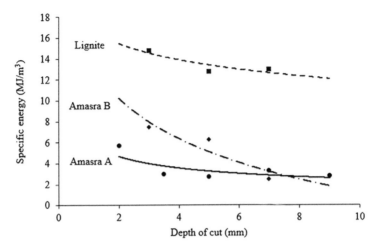

FIGURE 7.21
Variation of specific energy in unrelieved cutting mode for different coal specimens.

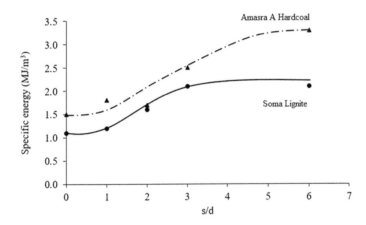

FIGURE 7.22
Variation of specific energy in relieved cutting mode for different coal specimens from Amasra A and Soma coalfields.

value for cutter spacing/depth of cut ratio of around one for coal specimens obtained from Amasra A and Soma coalfields. However, this optimum value is shown to be higher in lignite of Milten, which is around three, Figures 7.22 and 7.23.

The ratio of mean cutting force/mean normal force for simple chisel cutters was found to be less than for conical cutters, and it was around one for radial cutters at most depths of cut.

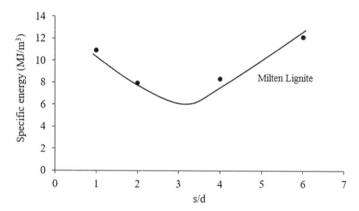

FIGURE 7.23
Variation of specific energy in relieved cutting mode for lignite of Milten.

7.7 Conclusive Remarks

Mechanization is an important strategy in the design and operation of modern mines. The objectives of mine mechanization include improved safety conditions, higher productivity, and a reduction in mining costs. A researcher or a mining engineer dealing with mechanical excavation of coal should first understand the basic principles of coal cutting mechanics affecting the coal cutting process. The efficiency of coal cutting machines depends on choosing right cutting tools with correct geometrical and operational parameters of cutters, such as depth of cut, the design of tool arrays, and their correct position in the cutting head. The inefficient torque and thrust may cause high specific energy, high amount of dust, etc. On the other hand, high forces acting on a drag tool may result in gross fracture damage to the tungsten carbide cutting tip or tool material and also damage to the machine components by exceeding the machine's torque and thrust capacities.

The works on coal cutting mechanics performed by Evans (1962, 1972, 1982, 1984a, 1984b) and Evans and Pomeroy (1966) are used to establish the basic principles of coal cutting, and these have been widely used in the past for efficient design of excavation machines such as shearers, continuous miners, and roadheaders. Since Evans' cutting theory is established for simple chisel cutters, a modification to the theory was made by Bilgin et al. (2012, 2014) for complex-shaped cutters by using some experimental coefficients already established by coal cutting experiments (Evans 1984a, 1984b) that demonstrated theoretically that tensile strength and compressive strength were dominant coal properties in point attack tools. Goktan (1990) suggested a modification on Evans' cutting theory for point attack tools, and he concluded that the force values obtained with this equation were close to previously published experimental values and could be of practical value, if confirmed by additional studies.

Coal cutting experiments realized by Roepke and Voltz (1983) in Bureau of Mines were used in this section to explain the basic cutting mechanism of conical and radial cutters and to compare the measured cutting force values with those predicted from Evans' cutting theory. It was found that the mean cutting force generated by 60° conical cutters and radial cutters were almost same for the depth of cut intervals between 12 mm and 25 mm, which was the interval where the most coal cutting machines work, and for 60° tool, the measured mean cutting force values were in good agreement with the predicted values within practical range of depth of cut (12 mm–15 mm) for most coal cutting machines. The comparison of measured coal cutting forces with predicted cutting force values from Evans' cutting theory for radial cutters showed that the measured mean cutting forces were in good agreement for almost all depth of cut changing from 3 mm to 25 mm.

Cutting force determines the need of torque and power of a cutting machine. Normal force plays an important role in sumping of the cutting head in the coal seam. There is not any theory to predict the normal force when cutting coal or rock with conical or radial cutters. This is predicted from cutting force/normal force ratios. The coal cutting experiments carried out by Roepke and Voltz (1983) showed that cutting force/normal force ratio increased linearly from around 1 for 3 mm depth of cut up to 2 for 25 mm depth of cut for 60° conical cutter. However, this ratio was less in 90° conical cutter changing from 0.7 to 1.2. With radial cutters, cutting force/normal force ratio changed from 1.3 to 4.9 for from shallow depth of 3 mm up to deeper depth of 25 mm.

The variation of cutting force with depth of cut for simple chisel cutters when cutting coal specimen from Amasra A Hard coalfield and Milten lignite field and comparison with predicted values from Evans' theory showed that for lignite specimen, the predicted cutting force was close to measured mean cutting force after a certain depth of cut, which was 15 mm in that case. However, for coal specimen obtained from Amasra A Hard coalfield and Milten lignite field, the measured mean cutting forces are almost twice as much as the predicted ones by Evans' theory. The variation of specific energy in unrelieved cutting mode for different coal specimens showed a decreasing trend with depth of cut as expected. For relieved cutting mode, specific energy showed an optimum value for cutter spacing/depth of cut ratio of around 1 for coal specimen obtained from Amasra A and Soma coalfields. However, this optimum value was shown to be higher in lignite of Milten, which was around 3. The ratio of mean cutting force/mean normal force for simple chisel cutters was found to be less than for conical cutters, and it was around 1 for radial cutters at most depths of cut.

References

Balci, C., Copur, H., Tumac, D., Avunduk, E., and Comakli, R. 2013. The cuttability of coal and coal measure rocks in relation to the performance prediction of road-headers in Amasra Coalfield B, for Hema AŞ. Project Report, Istanbul Technical University, Mining Engineering Department, 79 p.

Bilgin, N. 1977. Investigation into mechanical cutting characteristics of some medium and high strength rocks. PhD thesis, University of Newcastle Upon Tyne, UK, 332 p.

Bilgin, N., Balci, C., Copur, H., Tumac, D., and Avunduk, E. 2015. Cuttability of coal from the Soma coalfield in Turkey. *International Journal of Rock Mechanics and Mining Sciences*, 73:123–129.

Bilgin, N., Copur, H., and Balci, C. 2012. Effect of replacing disc cutters with chisel tools on performance of a TBM in difficult ground conditions. *Tunnelling and Underground Space Technology*, 27(1):41–51.

Bilgin, N., Copur, H., and Balci, C. 2014. *Mechanical Excavation in Mining and Civil Industries*. CRC Press/Taylor & Francis Group, Boca Raton, FL.

Bilgin, N., Copur, H., Balci, C., Avunduk, E., and Tumac, D. 2011. The cuttability and cavability of thick coal seam in Soma-Eynez coal field IR-75153 of TKI operated by IMBAT AS. Project report, Istanbul Technical University, 59 p.

Bilgin, N., Temizyurek, I., Copur, H., Balci, C., and Tumac, D. 2010. Cuttability characteristics of TTK Amasra thick seam and some comments on mechanized excavation. *Proceedings of the 17th Coal Congress of Turkey*, Editors. K. Colak, H. Aydın, June 2–4, Zonguldak, Turkey. ISBN: 978-9944-89-986-4, pp. 217–229.

Copur, H., Balci, C., Bilgin, N., Tumac, D., and Avunduk, E. 2012. Predicting cutting performance of chisel tools by using physical and mechanical properties of natural stones. *Proceedings of the European Rock Mechanics Symposium* (Eurock 2012, ISRM International Symposium), May 28–30, Stockholm, Sweden 14 p.

Copur, H., Bilgin, N., Balci, C., Tumac, D., and Avunduk, E. 2017. Effects of different cutting patterns and experimental conditions on the performance of a conical drag tool. *Rock Mechanics and Rock Engineering*, 50(6):1585–1609.

Eskikaya, S., Bilgin, N., Yazici, S., Acaroglu, O., and Arapoglu, B. 1995. The cuttability characteristics of lignite seams from Milten coalfield. Project Report, Istanbul Technical University, Mining Engineering Department, Turkey, 33 p.

Evans, I. 1962. A theory of the basic mechanics of coal ploughing. *Proceedings of the International Symposium on Mining Research*, University of Missouri Pergamon Press, Pergamon City, Turkey. V2, pp. 761–768.

Evans, I. 1972. Line spacing of picks for efficient cutting. *International Journal of Rock Mechanics and Mining Sciences & Geomechanics Abstracts*, 9(3):355–361.

Evans, I. 1982. Optimum line spacing for cutting picks. *The Mining Engineer*, 141:433–434.

Evans, I. 1984a. A theory of the cutting force for point attack picks. *International Journal of Mining Engineering*, 2(1):63–71.

Evans, I. 1984b. Basic mechanics of the point attack pick. *Colliery Guardian*, 232(5):189–193.

Evans, I. and Pomeroy, C. D. 1966. *The Strength, Fracture and Workability of Coal*. Pergamon Press, London, UK, 277 p.

Goktan, M. 1990. Effect of cutter pick rake angle on the failure pattern of high strength rocks. *Mineral Science and Technology*, 11(3):281–285.

https://www.gracesguide.co.uk/Anderson, _Boyes_and_Co, taken on April 2018, Grace's Guide to British Industrial History.

https://www.ncm.org.uk/downloads/195/Mine_machines_-_coal_cutting.pdf, taken on April 2018, Mine machines, coal cutting, IMH Group, National Coal Mining Museum for England.

https://www.scrapmonster.com/company/eickhoff-corporation/31681, taken on April 2018, Eickhoff Corporation.

Liu, S., Ji, H., Liu, X., and Jiang, H. 2017. Experimental research on wear of conical pick interacting with coal-rock. *Engineering Failure Analysis*, 74:172–187.

Mathey, M. 2015. Investigation into the mechanism of strength and failure in squat coal pillars in South Africa. PhD thesis, University of Witwatersrand, Johannesburg, 202 p.

Paschedag, U. 2014. Plow technology, history and the state of the industry. 2014 Caterpillar, Inc. Caterpillar Global Mining, 16 p.

Prokopenko, S., Sushko, A., and Kurzina, I. 2015. New design of cutters for coal mining machines. *IOP Conf. Series: Materials Science and Engineering*, 91(012058):1–8.

Roepke, W. W. and Voltz, J. I. 1983. *Coal-Cutting Forces and Primary Dust Generation Using Tadial Gage Cutters*. Report of Investigations: USA Bureau of Mines, Avondale, MD. RI 8800:19–24p.

Roxborough, F. F. 1973. Cutting rock with picks. *The Mining Engineer*, 132(153):445–454.

Roxborough, F. F. 1985. Research in mechanical excavation, progress and prospects. *Proceedings of Rapid Excavation and Tunnelling Conference*, Las Vegas, Nevada, pp. 225–244.

Roxborough, F. F. and Liu, Z. C. 1995. Theoretical considerations on pick shape in rock and coal cutting. *Proceedings of the 6th Underground Operator's Conference*, Kalgoorlie, WA, Australia, pp. 189–193.

Roxborough, F. F. and Rispin, A. 1973a. The mechanical cutting characteristics of the lower chalk. *Tunnels and Tunnelling*, 5:45–67.

Roxborough, F. F. and Rispin, A. 1973b. A laboratory investigation into the application of picks for mechanized tunnel boring in the lower chalk. *The Mining Engineer*, 133(1):1–13.

Shao, W., Li, X., Sun, Y., and Huang, H. 2017. Parametric study of rock cutting with smart/cut picks. *Tunnelling and Underground Space Technology*, 61:134–144.

Stack, B. 1995. *Encyclopedia of Tunneling, Mining and Tunneling Equipment*. Muden Publishing, Hobart, Tasmania, Vol. 2. ISBN 0 9587 11 3 8.

8

Coal Properties Effecting Coal Cuttability

8.1 Introduction

Coal plays an important role in power generation, with around 40% of the world's electricity coming from coal. It is believed that its importance will continue for decades (IEA 2013). Almost half of the worldwide underground coal production comes from longwall mining methods using shearers, ploughs, and continuous miners. However, it is essential to determine cuttability of coals before selecting the most appropriate cutting machine in a mechanized system for efficient coal production. Bearing in mind that cuttability and strength properties of coal are well defined in a work published by Evans and Pomeroy (1966). A tensile coal breakage theory was developed and has been used for many years for designing coal excavating machines. However, since then further research studies enlightened some more aspects of the relation between coal properties and cuttability of coal. This chapter is aimed to summarize some works realized in this respect.

8.2 Petrological and Mechanical Properties Effecting Coal Cuttability

Falcon (1978) attempted to review the application of coal petrology to classify coal cuttability in South Africa. He emphasized that in terms of horsepower required to break the lithotypes with continuous miners, fusain required least power, vitrain required twice as much as fusain, clarain three times as much, and durain seven times as much.

MacGregor (1983) pointed out that certain variables, such as percentage volatiles and some vitrinite, show some degree of correlation with certain mechanical properties and cuttability of South African coals. The study also endorsed the view stated by Mackowsky (1967) that the breaking properties of coal did not depend on the hardness, which is a characteristic of a homogenous material, but on the strength which characterizes the mechanical behavior of the heterogeneous substances (the term heterogeneous especially refers to cleat frequency).

MacGregor and Baker (1985) discussed the possible use of petrological data and proximate analysis in the prediction of cutting forces for coal, pointing out that the significant predictors were dependent on coal properties.

Szwilski (1985, 1987) stated that the cuttability was influenced largely by the structural and physical properties of the coal seam. The correlation between the rank, ash content, coal strength, and Schmidt hammer hardness values were used to classify the coal according to the cuttability or mineability characteristics.

An investigation into cuttability of coal seams and recording performance of a mechanical plough in Aegean lignite mine, Turkey, showed that uniaxial compressive strength, point load index, Schmidt hammer rebound value, National coal board (NCB) cone indenter hardness, and impact strength index values may be used in a strength classification of coal seams for cuttability assessment in this mine as shown in Table 8.1. However, it is strongly emphasized that coal is a complex material, for any strength classification system, one should be careful not to generalize a particular set of data, bearing in mind that any special application may carry its own characteristics (Bilgin et al. 1992, Bilgin and Phillips 1994).

TABLE 8.1

Cuttability Classification of Coal Seams in Aegean Lignite Mine, Turkey for a Mechanical Plough, Rearranged After

Strength Properties	Easy to Cut	Medium to Cut	Difficult to Cut	Very Difficult to Cut
Compressive strength, (kg/cm^2)	<120	120–170	170–250	>250
Point load index, I_{50} (kg/cm^2)	<3.5	3.5–4.5	4.6–11.0	>11
Cone indenter	<1.5	1.5–1.8	1.8–2.6	>2.6
Schmidt hammer rebound value, N type	<20	20–27	27–42	>42
Impact strength index	<45	45–58	58–72	>72

Source: Bilgin, N. et al., The cuttability classification of coal seams and an example to a mechanical plough application in E.L.I. Darkale Coal Mine, *Proceedings of the 8th Coal Congress of Turkey*, Zonguldak, Turkey, pp. 31–53, 1992; Bilgin, N. and Phillips, H.R., Mechanical properties of coal, in *Coal, Resources, Properties Utilization, Pollution*, Kural, O., Ed., Istanbul, Turkey, 1994.

8.3 Specific Energy

Specific energy is defined as the amount of work required to break a unit volume or mass of rock/coal and used to predict the performance of mechanical miners. It can be obtained from full-scale rock cutting experiments, in-situ tests, the energy consumed by the cutting machines, or by predicting empirically from rock/coal mechanical properties (Copur et al. 2001, Balci et al. 2004, Balci and Bilgin 2007, Dogruoz and Bolukbasi 2017).

8.3.1 In-Situ Specific Energy

In the past, different testing devices were developed for determination of the in-situ cuttability of rock/coal. However, as Singh (1987) stated, these methods developed in the past certainly provided very valuable information for in-situ cuttability of coal, but only a few of them closely simulated the operation of a cutter at the coal face. However, in-situ specific energy values changing between 52.5 MJ/m^3 and 83.9 MJ/m^3 obtained by Singh (1987) for UTAH coals using a portable testing instrument are still beyond the values of field specific energy values changing between 0.82 MJ/m^3 and 4.80 MJ/m^3 obtained from power consumption of shearers used in Indian coals (Singh et al. 1995). Field specific energy of 0.82 MJ/m^3 was obtained in a coal seam having Schmidt hammer rebound value of 23 and cleat frequency of 15 (number per meter), field specific energy value of 4.8 MJ/m^3 was obtained in a coal seam having Schmidt hammer rebound value of 52 and cleat frequency of 17 (Singh et al. 1995). As Myszkowski and Paschedag (2013) emphasized that the specific energy of shearers varied between 0.7 MJ/m^3 and almost 10 MJ/m^3, although in a variety of cases it did not exceed 5 MJ/m^3. Plow systems are characterized by a specific energy ranging from 1.0 MJ/m^3 to nearly 10 MJ/m^3, but in most plow faces the specific energy does not go beyond 5 MJ/m^3. Based upon those numbers, it can be stated that the specific energy values for both extraction systems are comparable.

8.3.2 Laboratory Specific Energy

Laboratory specific energy is the most reliable method for defining cuttability of coal; it is found dividing the mean cutting force in kilonewton (kN) to yield in m^3/km to find the specific energy in MJ/m^3. However, it is important to note that specific energy is not only represented by the strength characteristics of coal, but also represented by the cutting parameters like cleat orientation, depth of cut, rake angle, clearance angle, the geometry of cutting tool, etc. (Roxborough and Hagen 2010).

Figures 8.1 through 8.3 published by Roxborough and Hagen (2010) on the effect of rake angle, depth of cut, and cleat orientation on specific energy in Bulli coal seam clearly emphasize the need of standardization of specific energy for a

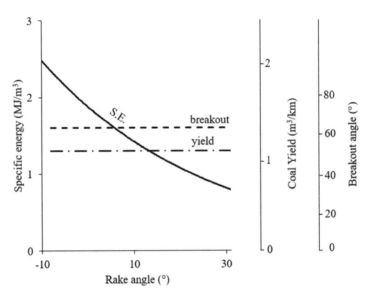

FIGURE 8.1
Effect of rake angle on specific energy in Bulli coal seam. (From Roxborough, F.F. and Hagen, P., Elements of machine mining, module reader. *Mining Education Australia (MEA)*. Minerals Council of Australia, 88, 2010.)

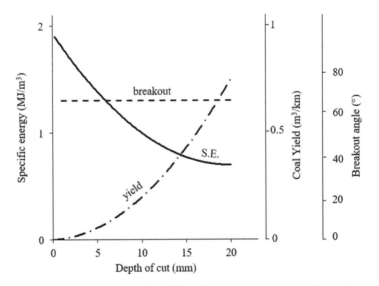

FIGURE 8.2
Effect of depth of cut on specific energy in Bulli coal seam. (From Roxborough, F.F. and Hagen, P., Elements of machine mining, module reader. *Mining Education Australia (MEA)*. Minerals Council of Australia, 88, 2010.)

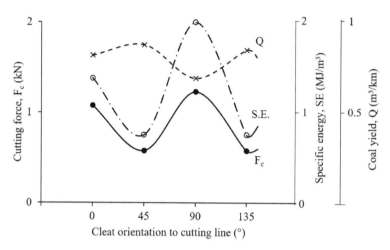

FIGURE 8.3
Effect of cleat orientation on specific energy in Bulli coal seam. (From Roxborough, F.F. and Hagen, P., Elements of machine mining, module reader. *Mining Education Australia (MEA)*. Minerals Council of Australia, 88, 2010.)

comparative study by keeping the cutting parameters constant. For this, a standard cutting test using a chisel cutter having width of 12.7 mm, rake angle of −5°, and 5 mm of depth of cut was suggested by Fowell and McFeat-Smith (1976) and Roxborough and Sen (1986). This test has been used by the authors for a long time in Istanbul Technical University, Mining Engineering Laboratories.

Eight different coal samples taken from different levels of the coal seams were subjected to cuttability tests by the authors of this book in order to see the variability of coal cuttability along the seam (Bilgin et al. 2015). A laboratory cutting set, similar to the one developed in the University of Newcastle Upon Tyne, United Kingdom as described by Fowell and McFeat-Smith (1976) was used for cuttability determination of the coal samples. A chisel cutter having a width of 12.7 mm and rake angle of (−5°) was used in the experiments at a cutting depth of 5 mm. The three orthogonal force components acting on the tool were recorded with a data acquisition system sampled at 1000 times every second. The experiments were executed parallel to bedding planes in unrelieved cutting mode. Specific energy is one of the most important parameters in determining the cuttability of coals since the production rate (PR) may be determined for any excavating machine using specific energy as given in Equation 8.1.

$$PR = k \cdot P/SE \tag{8.1}$$

where PR is production rate in m³/h, P is the cutting power of the mechanical cutting device or coal cutting shearer in kW, k is the energy transfer ratio from cutting machine to the coal face which is changing between 0.8 and 0.9 for shearers and ploughs, and SE is the specific energy in kWh/m³.

Specific energy provides also for selection of the mining system. In most cases, shearers are used in harder coals compared to plows, in other words, shearers are preferred in coals having higher specific energy values (Bilgin et al. 2014a).

Copur et al. (2001) and Balci et al. (2004) showed typical examples of how specific energy concept is used for selecting a rapid excavation system such as roadheader and predicting its performance.

There is always an optimum specific energy value for a given cutter spacing/depth of cut ratio allowing also the most efficient cutting head design. The cutter spacing is fixed in a predetermined excavating machine, however, the depth of cut is an operational parameter, which can be controlled during the excavation. This concept permits to work in optimum conditions by adjusting carefully the depth of cut using proper thrust force and rotational speed.

In the research work described above, the specific energy measured in the laboratory was for one depth of cut and one tool geometry under standardized conditions. However, a detailed research program was carried out recently, and it was reported that specific energy values measured under standardized laboratory cutting conditions were closely related to optimum specific energy values measured in relieved cutting (Balci and Bilgin 2007), where there was interaction between cutting grooves. A ratio of optimum specific energy to standard specific energy value was found to be around 0.7.

The relationship between specific energy and Schmidt hammer rebound values is given in Figure 8.4, as it is seen clearly from this figure, specific energy increases linearly with Schmidt hammer rebound values suggesting that Schmidt hammer test may serve as a valuable tool in estimating the cuttability of coal seams. These results support the findings reported by Singh et al. (2002).

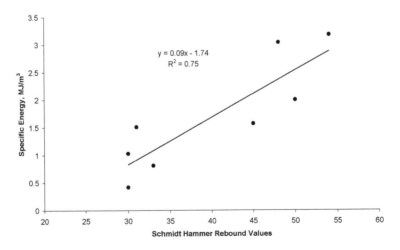

FIGURE 8.4

Relationship between Schmidt hammer hardness (type N) and specific energy. (From Bilgin, M., et al., *International Journal of Rock Mechanics and Mining Sciences*, 73, 123–129, 2015.)

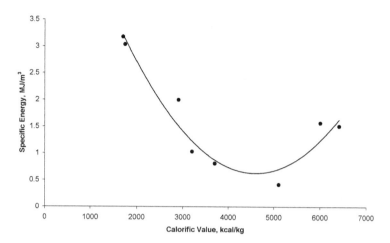

FIGURE 8.5
Relationship between calorific value and specific energy. (From Bilgin, M., et al., *International Journal of Rock Mechanics and Mining Sciences*, 73, 123–129, 2015.)

As seen in Figure 8.5, specific energy shows an optimum value for a given calorific value of around 4,000 kcal/kg–5,000 kcal/kg.

The variation of specific energy with volatile matter, fixed carbon, and ash content show, as in Figures 8.6 through 8.8, the same trend as calorific value, minimum or optimum values are obtained for volatile matter of 35%, fixed carbon of 35%, and ash content of 20%.

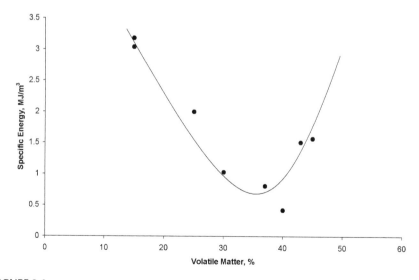

FIGURE 8.6
Relationship between volatile matter and specific energy. (From Bilgin, M., et al., *International Journal of Rock Mechanics and Mining Sciences*, 73, 123–129, 2015.)

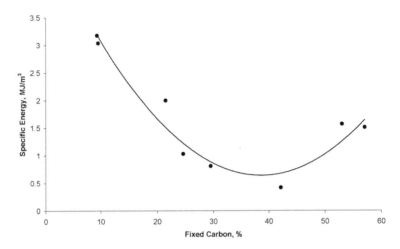

FIGURE 8.7
Relationship between fixed carbon and specific energy. (From Bilgin, M., et al., *International Journal of Rock Mechanics and Mining Sciences*, 73, 123–129, 2015.)

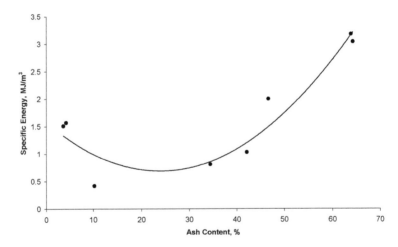

FIGURE 8.8
Relationship between ash content and specific energy. (From Bilgin, M., et al., *International Journal of Rock Mechanics and Mining Sciences*, 73, 123–129, 2015.)

As a summary of the works mentioned above, a new excavability classification of coal based on Schmidt hammer rebound values, coal properties, and specific energy is given in Table 8.2.

It is also worth mentioning that the work was done by Singh et al. (2002) to determine bit forces encountered in Indian coal types (having a wide range of compressive strength and hardness), with a view to enabling the design engineers to estimate the performance of excavating machines in different

TABLE 8.2

New Excavability Classification of Coal Based on Schmidt Hammer Rebound Values, Coal Properties and Specific Energy

Excavability Classification	SH₃	SHRF₃	E.C. (kg/m³)	UCS (MPa)	C.V. (%)	V.M. (%)	F.C. (%)	A.C. (%)	SE (MJ/m³)
Easy	<48	30–59	<0.35	<18	4000–5000	30–40	30–45	20–30	1.0–1.5
Moderate	48–54	20–30	0.35–0.40	18–27	4000–2500	40–45, 30–25	30–20, 45–55	<20, 30–40	1.2–2.0
Difficult	54–60	14–20	0.40–0.50	27–70	2500–2000 >5000	25–15	20–15	40–50	2.0–4.0
Very difficult	>60	<20	>50	>70	<2000	<15	<15	>50	>4.0

SH₃: Mean of the highest three values within one set of Schmidt hammer readings measured in the field, SHRF₃: Schmidt hammer reduction factor estimated by using the mean values of the highest three Schmidt hammer readings (SH₃), E.C.: explosive consumption used to ease the workability of coal, UCS: uniaxial compressive strength of coal, C.V.: calorific value of coal, V.M.: volatile matter of coal, F.C.: fixed carbon content of coal, A.C.: ash content of coal, and SE: specific energy obtained from unrelieved cutting tests under standardized conditions.

coal seams. Tests were carried out on an electro-hydraulic coal plough rig designed and developed at the Central Mining Research Institute, Dhanbad. Bits of different geometries were tested on 12 coal types with uniaxial compressive strength ranging from 15 MPa to 55 MPa. Cutting tests were performed at depths up to 9 mm and at speeds up to 30.8 cm/s. The effect of varying the depth of cut was determined and specific energies calculated. For the coal tested, relationships were established between cutting force and depth of cut, specific energy and depth of cut. Finally, relationships were established between specific energy and coal mechanical properties.

8.4 Breakout Angle and Coal Workability

In the last years (2011, 2012), in-situ test devices for determining and evaluating the cuttability of coal materials have been developed and used in Poland and in some other countries (Bialy 2015). These devices reproduce the work of a coal plough and a longwall shearer, respectively. The first of the instruments was invented at the Central Mining Institute in Katowice, whereas the other one was developed at the Silesian University of Technology in Production Engineering Institute. The instruments were based on state-of-the-art studies in terms of construction as well as measurement and recording. The instruments in question have Atex I M2 Ex and I Mb certificate, allowing their work as devices intended for use in explosive atmospheres in underground coal conditions (Bialy 2015). Basically, chisel and conical cutters are used and cutting forces and breakout angles as defined in Figure 8.9 are used as

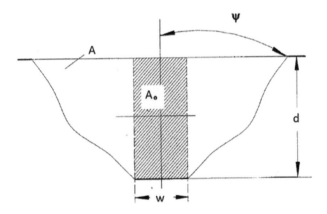

FIGURE 8.9
Illustration of breakout angle. (From Bialy, W., *Management Systems in Production Engineering*, 20, 202–209, 2015.)

TABLE 8.3

Classification of Coal Workability by Bialy Based on In Situ Coal Cutting Tests

Breakout Angle, ψ	Strength Classification	Workability Rate, Aψ (kN/cm)	Workability Classification
$\psi > 70°$	Brittle	$A_\psi \leq 1.80$	Easily workable
$40° < \psi \leq 70°$	Hard	$1.81 < A_\psi \leq 3.00$	Medium workable
$\psi \leq 40°$	Very hard	$A_\psi > 3.01$	Hardly workable

Source: Bialy, W., *Management Systems in Production Engineering*, 20, 202–209, 2015.

a criterion for workability or cuttability of coal. Depth of cut is kept at 1 cm and the width of chisel cutters as 2 cm in most cases. Workability rate in (kN/cm) as denoted as A_ψ and breakout angle ψ is used for workability/cuttability classification of coal as given in Table 8.3. Workability rate is calculated using the concept of cutting resistance in relation to cutting force (FC), groove area (A), cutter area (A_o), breakout angle (ψ), and depth of cut (d) as shown in Figure 8.9. The expression used to calculate workability rate is given in Equation 8.2.

$$A_\psi = \frac{FC}{d \cdot \left(1 + (d/w) \cdot \tan\psi\right)} \tag{8.2}$$

However, it should be noted that breakout angle at 1 cm depth of cut should be considered in Table 8.3. Breakout angle changes with depth of cut, as the authors noticed in testing coal specimen from Amasra coalfield (B)-Hema, Turkey, which was found 82° for depth of cut 0.3 cm, 80° for depth of cut 0.5 cm, and 59° for depth of cut of 0.7 cm (Bilgin et al. 2010). For 1 cm depth of cut, the breakout angle for Zonguldak coalfield (Turkey) is found to change between 60° and 65° with specific energy changing between 5.7 MJ/m³ and 14 MJ/m³ (Bilgin et al. 1994b). According to classification given by Bialy (2015) in Table 8.3, the hard coal from Zonguldak may be classified as medium workable coal.

8.5 Conclusive Remarks

The basic objectives of these research studies were to find the relationships between cutter forces, specific energy, mechanical strength of coals, and cleat frequencies. Classifications of the coal seams according to the workability were based on these findings. However, this research differs from the previous ones in a sense that the cuttability or specific energy is found to be related to coal properties such as calorific value, volatile matter, fixed carbon, and ash content. Specific energy shows an optimum value for a given calorific value of around 4000 kcal/kg–5000 kcal/kg. The variation of specific energy with volatile matter, fixed carbon, and ash content show the same trend. Minimum or optimum values are obtained for volatile matter at 35%, fixed carbon at 35%, and ash content at 20%. These findings show that measuring specific energy during coal production with shearers, ploughs, and continuous miners may serve as a basis to classify and store the produced coal according to the coal quality, hence, feeding the power station with blended coal production to a desired and designed coal quality value for which the power plant is designed. A new workability/cuttability classification of the coal has been suggested using the findings of the authors of this book. It is also worth noting that the workability/cuttability classification given by Bialy (2002, 2015) is in good accordance with the cutting test results obtained by the authors of this book.

References

Balci, C. and Bilgin, N. 2007. Correlative study of linear small and full scale rock cutting tests to select mechanized excavation machines. *International Journal of Rock Mechanics and Mining Science,* 44(3):468–476.

Balci, C., Demircin, M. A., Copur, H., and Tuncdemir, H. 2004. Estimation of optimum specific energy based on rock properties for assessment of roadheader performance. *Journal of the Southern African Institute of Mining and Metallurgy,* 104(11):633–642.

Bialy, W. 2002. The side-crumble angle ψ of coal and the energy consumption of the mining process as a function of the vertical component σz of exploitation pressure. *Archiwum Gornictwa,* 47(3):361–384.

Bialy, W. 2015. Innovative solutions applied in tools for determining coal mechanical properties. *Management Systems in Production Engineering,* 20(4):202–209.

Bilgin, N., Balci, C., Copur, H., Tumac, D., and Avunduk, E. 2015. Cuttability of coal from the Soma coalfield in Turkey. *International Journal of Rock Mechanics and Mining Sciences,* 73:123–129.

Bilgin, N., Copur, H., and Balci, C. 2014a. *Mechanical Excavation in Mining and Civil Industries.* CRC Press/Taylor & Francis Group, Boca Raton, FL.

Bilgin, N., Copur, H., Balci, C., Tumac, D., Avunduk, E., and Comakli, R. 2014b. Study on the full scale laboratory cutting experiments carried out to determine the cuttability characteristics of the coal from Zonguldak Coal Field. Project Report, Istanbul Technical University, Mining Engineering Department, Turkey, 49 p.

Bilgin, N. and Phillips, H. R. 1994. Mechanical properties of coal. In: *Coal, Resources, Properties Utilization, Pollution.* Editor. O. Kural, Istanbul, Turkey, Mining Faculty, Istanbul Technical University. ISBN: 975-95701-1-4.

Bilgin, N., Phillips, H. R., and Yavuz, N. 1992. The cuttability classification of coal seams and an example to a mechanical plough application in E.L.I. Darkale Coal Mine. *Proceedings of the 8th Coal Congress of Turkey,* Zonguldak, Turkey, pp. 31–53.

Bilgin, N., Temizyurek, I., Copur, H., Balci, C., and Tumac, D. 2010. Cuttability characteristics of TTK Amasra thick seam and some comments on mechanized excavation. *Proceedings of the 17th Coal Congress of Turkey*, Editors. K. Colak, H. Aydın, June 2–4, Timmob Maden Mühendisleri Odasi Zonguldak Şubes, Zonguldak-Turkey. ISBN: 978-9944-89-986-4, pp. 217–229.

Copur, H., Tuncdemir, H., Bilgin, N., and Dincer, T. 2001. Specific energy as a criterion for the use of rapid excavation systems in Turkish Mines. *Transactions of the Institutions of Mining and Metallurgy: Section A*, 110(3):149–157.

Dogruoz, C. and Bolukbasi, N. 2017. Effect of cone indenter hardness on specific energy of rock cutting. *Suleyman Demirel University Journal of Natural and Applied Sciences*, 21(3):743–748.

Evans, I. and Pomeroy, C. D. 1966. *The Strength, Fracture and Workability of Coal.* Pergamon Press, London, UK.

Falcon, R. M. S. 1978. Coal in South Africa part III, summary and proposals-the fundamental approach to the characterization and rationalization of South Africa's coal. *Miner Science Engineering*, 10(2):130–152.

Fowell, R. J. and McFeat-Smith, I. 1976. Factors influencing the cutting performance of a selective tunnelling machine. *Proceedings of the International Symposium IMM (Tunelling'76)*, IMM, London, UK, pp. 301–309.

IEA. 2013. *Key World Energy Statistics*, International Energy Agency, Paris, France, 82 p. Online: www.iea.org/publications/freepublications/publication/Key World2013.pdf.

MacGregor, I. M. 1983. Preliminary results on the relationship of coal petrology to coal cuttability in some South African coals. *The Geological Society South Africa*, 7:117–128.

MacGregor, I. M. and Baker, D. R. 1985. A preliminary investigation of predictors of the cutting forces for some South African coals. *Journal of the Southern African Institute of Mining and Metallurgy*, 85(8):259–272.

Mackowsky, M. T. 1967. Progress on coal petrology. *Proceeding of the Symposium on the Science and the Technology of Coal*, Mines Branch, Department of Energy, Mines and Resources, Ottawa, Canada, pp. 60–78.

Myszkowski, M. and Paschedag, U. 2013. *Longwall Mining in Seams of Medium Thickness Comparison of Plow and Shearer Performance Under Comparable Conditions.* Caterpillar, 51 p.

Roxborough, F. F. and Hagen, P. 2010. Elements of machine mining, module reader. *Mining Education Australia (MEA)*. Minerals Council of Australia, 88 p.

Roxborough, F. F. and Sen, G. L. 1986. Breaking coal and rock, Australian coal mining practice. *The Australian Institute of Mining and Metallurgy*, 12:130–147.

Singh, R., Singh, A. K., and Mandal, P. K. 2002. Cuttability of coal seams with igneous intrusion. *Engineering Geology*, 67(1–2):127–137.

Singh, R., Sing, J. K., Singh, T. N., and Dhar, B. B. 1995. Cuttability assessment of hard coal seams. *Geotechnical and Geological Engineering*, 13(2):63–78.

Singh, S. P. 1987. Criterion for the assessment of the cuttability of coal. In: *Underground Mining Methods and Technology*, Editors A.B. Szwilski and M.J. Richards. Elsevier Science Publishers, B.V. Amsterdam, the Netherlands, 225–239.

Szwilski, A. B. 1985. Relation between the structural and physical properties of coal. *Mining Science and Technology*, 2:181–189.

Szwilski, A. B. 1987. Evaluation of structural properties of coal seams. *Mining Engineering*, 2:115.

9

Workability of Coal, Drilling and Digging/Breaking the Coal

9.1 Introduction

Workability of coal is a general name given to the cuttability, grindability, drillability, and diggability/excavability of coal. Cuttability and grindability of coal were discussed in Chapters 7 and 8. Directional drilling and diggability/excavability of coal are two other important subjects of underground and surface mining, so they will be discussed separately in this chapter.

9.2 Drilling in Coal Mining

Drilling is an important component of the coal mining and used for multiple purposes including ore body characterization (with electronic geo-sensing technologies and sampling), surface and underground blast hole drilling, underground roof and wall bolting and cabling, dewatering, and especially directional drilling for methane drainage.

9.2.1 In-Mine Directional Drilling

In-mine directional drilling provides the coal mining industry with effective and practical options for methane drainage and exploration in advance of mining. Measurement and imaging provide for coal quality information while drilling, which increases the value of directional drilling. Geomechanical properties of coal seams are relevant for assessing drillability and borehole stability. Frequency and type of geologic discontinuities, including stress orientations, affect borehole stability and dictate permeability anisotropy. These data may help designers to help planning borehole locations and orientations to maximize methane drainage. In-situ gas contents, natural fracture and cleat permeability, and desorption characteristics determine times required for in-seam methane drainage and borehole spacing.

In coal with high methane content, as Brunner and Schwoebel (2018) stated, in-seam boreholes may have a significant effect on reducing ventilation requirements and increasing daily production rates. A typical example is longwall mining operations at Monclova Mining Company, in northern Mexico. High cleat and natural fracture permeability of coal seams, in excess of 30 mD, helped in-seam boreholes to apply a methane emission program (1000 m^3 per day) very efficiently allowing significant increase in coal production. Methane emissions into the entry decreased by 30% in two months, enabling the mine to reduce ventilation requirements and to increase mining advance rates by 78%.

9.2.2 Directional Drilling Technology for Gas Drainage and Exploration

Unconventional gas projects showed a tremendous increase worldwide in recent years. Table 9.1 shows CBM (coalbed methane), shale gas, and tight sand gas productions by the major gas procuring countries (https://www.iea.org/ugforum/ugd/). Figure 9.1 illustrates the occurrence of coalbed methane, shale gas, and tight sand gas (https://geology.com/energy/shale-gas/2018). Shale gas refers to natural gas that is trapped within shale formations. Shales are fine-grained sedimentary rocks that can be rich in sources of petroleum and natural gas.

As seen in Table 9.1, China had no CBM production in 2000, but in 2014, 13.0 billion m^3 was extracted from its extensive underground coal reserves. In Canada, CBM production was 0.584 Bm^3 in 2000 and increased to 7.18 Bm^3 in 2014. These amounts are small compared with United States production in 2014 which was 37.095 Bm^3. However, all this production is significant because it comes from an energy resource that was barely utilized

TABLE 9.1

Unconventional Gas Productions by Country

Country	CBM 2000 (Bm³)	CBM 2014 (Bm³)	Shale Gas 2000 (Bm³)	Shale Gas 2014 (Bm³)	Tight Gas 2000 (Bm³)	Tight Gas 2014 (Bm³)
USA	39.049	37.095	23.049	378.811	94.245	127.709
China	0	13.00	0	4.47	0	13.53
Canada	0.584	7.18	0	5.935	32.12	72.927
Australia	0.985	7.051	0	0	0	0
Russia	0	0.5	0	0	8.077	20.768
Germany	0.91	0.902	0	0	1.023	0.407
Argentina	0	0.305	0	0.305	2.427	2.207

Source: International Energy Agency (IEA), https://www.iea.org/ugforum/ugd/Unconventional gas production database, 2018.

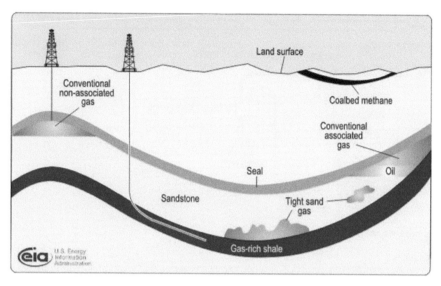

FIGURE 9.1
Illustration of the occurrence of coalbed methane, shale gas, and tight gas. (https://geology. com/energy/shale-gas/ uploaded on 2018.)

before 1985. It is important to note that United States shale gas production in 2000 increased from 23.048 Bm^3 to 378.811 Bm^3 in 2014. Acceptance of this unconventional resource as an alternative natural gas supply is evidenced by the level of investment capital being expended worldwide.

Recently, directional, horizontal drilling and hydraulic fracturing became one of the main subjects of the research studies directly related to the production of CBM. Typical example is China, 3267 papers on CBM were published by 245 Chinese journals from 2003 to 2013 (Yong and Jianping 2015).

In the past 30 years, in-mine borehole steering equipment has been developed from the single shot camera survey systems to the advanced directional drill monitor, which enables borehole survey measurements during drilling operations. Discontinuities encountered during in-seam directional drilling, such as faults, folds, and igneous intrusions, can be monitored by drilling fluid pressures, changes in thrust, vibration, rate of penetration, and inspection of cuttings. Typical in-seam directional equipment is seen in Figure 9.2 (Thomson 1997), and a typical surface to in-seam horizontal drainage drilling application is illustrated in Figure 9.3 (Wang et al. 2011).

Over the past decade, the combination of horizontal drilling and hydraulic fracturing has allowed access to large volumes of shale gas that were previously uneconomical to produce. The production of natural gas from shale formations has reactivated the natural gas industry in the United States.

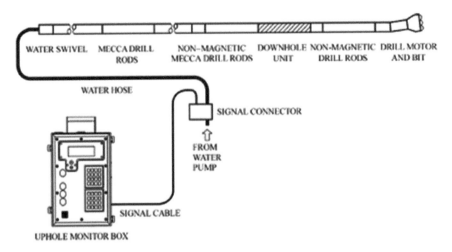

FIGURE 9.2
Directional drilling downhole equipment. (From Thomson, S., Directional drilling for safety in coal mining, in: Doyle, Moloney, Rogis & Sheldon, editors. *Proceedings of the Symposium on Safety in Mines: The Role of Geology,* November 24–25, pp. 77–84, 1997.)

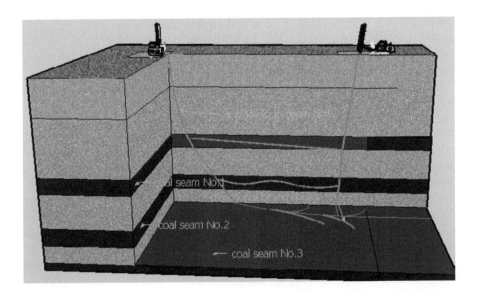

FIGURE 9.3
A typical surface to inseam horizontal drainage drilling application. (From Wang, F. et al., *First International Symposium on Mine Safety Science and Engineering, Procedia Engineering,* 26, 25–36, 2011.)

9.3 Digging/Breaking the Coal

About 60% of global coal production comes from underground mines. However, surface mining is more common in several major coal producing countries. For example, around 80% of coal production in Australia comes from surface mining, and, in the United States, about 30% of production is from surface mining (Fuginski 2012). However, surface mining operations are steadily increasing in recent years to win coal and minerals from greater depths due to the improved design of powerful mechanical excavators, such as bucket wheel excavators (BWEs), hydraulic excavators, walking draglines, surface continuous miners, stripping, and loading shovels. The selection of an excavator is a prime importance because it largely determines the other equipment required and the mode of operation. The performance of an excavator mainly depends on: (a) the design parameters of the cutting head, (b) power of the machine, (c) the intact strength and the abrasivity properties of the coal, (d) the competence of the ground as a whole, and (e) the new properties of the broken coal (Bilgin and Balci 2005).

9.3.1 Bucket Wheel Excavators

General view of a bucket wheel excavator is given in Figure 9.4. Although its application is extending to harder formations, the BWE is the most effective machine for mining large outputs in weak unconsolidated rock. Bucket wheel belt conveyor systems can achieve high production rates. It is reported that with a big capacity BWE, it is possible to obtain production rates of 10,000 m³/day–200,000 m³/day. Because the BWEs have less operating flexibility than shovel trucks, a detailed study must be carried out before selecting the machine (Kennedy 1990). Continuous excavators are usually rated in terms of theoretical output as given by Singh (1993):

$$Q = \frac{60 \cdot F \cdot s}{\text{Swelling Factor}} \tag{9.1}$$

where Q is the theoretical output in m³/h, F is the capacity of a single bucket, s is the number of bucket discharges per minute, and the swell factor is that of materials being excavated. Wheel radius of BWE changes usually between 2.2 m and 10.8 m, having theoretical outputs between ~720 and 12,470 m³/h and installed wheel drive power between 110 and 3 × 840 kW (Rasper 1975).

There are many advantages of using BWEs:

1. Excavation is continuous and there is not cycle times;
2. For a given output, a BWE is actually smaller than a dragline and shovel;

FIGURE 9.4
A bucket wheel excavator working in Libous coal mine. (From Sladkova, D., *Process Automation, Management Systems in Production Engineering*, 2, 3–7, 2011.)

3. Belt conveyors or trucks may be used for loading the excavated material;

4. It has no shock loading and has lower power requirements; and

5. It can be designed with ground bearing pressure as low as 75.8 kPa, but a reasonable design is 137.9 kPa.

Disadvantages of a BWE may be cited as:

1. It has high initial cost for a given production rate;

2. It cannot handle consolidated hard materials;

3. It has low availability since its machinery is highly complicated; and

4. Thickness of coal would have to be at least 0.7 times the diameter of the wheel (Stefanko 1983).

Hamrikova and Jurman (2011) emphasized strongly that supplied energy had to be utilized more effectively, and the operational reliability of machines had to be improved. Property characterizing the rock-machine interaction

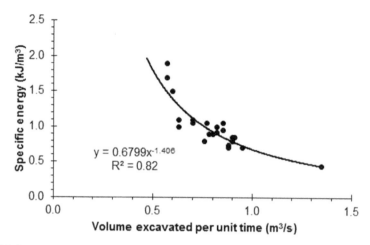

FIGURE 9.5
Relationship between specific energy and volume excavated per unit time. (After Hamrikova, R. and Jurman, J., *Management Systems in Production Engineering*, 2, 8–11, 2011.)

is called digging resistance or specific energy and, as seen in Figure 9.5, there is a minimum specific energy for a given volume of excavated material obtained in unit time. Accepting the quantity of minimum specific energy by volume as a criterion for evaluating the digging resistance made it possible to form a new and more objective opinion on the impacts of particular factors affecting the final effect of the mining. To eliminate a negative effect of the operator of the excavator, the installation of an information system is necessary. In virtue of the minimization of specific energy by volume, the system will make it possible to determine needful dimensions of the cut and, thus, will enable the optimum regulation of excavator efficiency, naturally, all above mentioned being in accordance with other control and protective elements of the excavator.

9.3.2 Hydraulic Excavators

General view of a bucket hydraulic excavator is given in Figure 9.6. Although hydraulic excavator was introduced in United States as a backhoe, it was developed in Europe. Most hydraulic excavators are diesel, powered electrical versions are also available. Its unique application of power, bucket agility, speed, and ease of operations exceeds conventional rope (cable) shovels in many instances. It is being increasingly used in overburden up to 9.1 m, especially with the various hauling techniques. In estimating operating costs, availabilities of 90% are usually used. In calculating ownership costs, it is considered as a medium life machine from 8 years to 10 years.

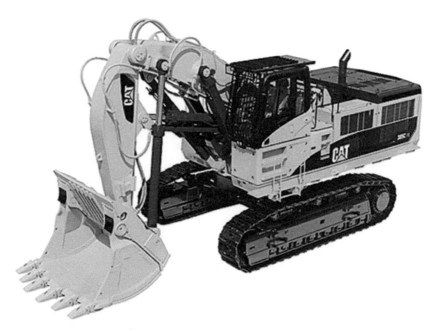

FIGURE 9.6
A hydraulic excavator, Cat 385C FS. (Courtesy of Caterpillar, Peoria, IL.)

9.3.3 Power—Cable, Electrical—Shovels

Typical view of a power shovel is seen in Figure 9.7. The power shovel is designed for stripping or loading. They have greatest application in handling tight or poorly fractured overburden because of its crowding action, which allows a higher breakout force to be applied. In shovel applications, cycle times are normally low, a complete cycle for 90 dumping is generally 60 seconds. Their capacities range from 10 m³ to 140 m³ with boom lengths up to 72 m. Power shovels play an important role in certain strip mining because they strip off the overburden and dig and load the coal. Combination of shovels and trucks makes possible to move the overburden farther than reach of any dragline.

An extensive field research program for diggability classification of different type and size of electrical shovels at TKI's (Turkish Coal Enterprises) surface coal mines was carried out by Ceylanoglu et al. (1994). Specific digging energy was found most suitable for diggability classification as suggested in Table 9.2. Ceylanoglu et al. (1994) proved also that normalized specific energy might be predicted from uniaxial compressive strength, tensile strength, Schmidt hardness, seismic P wave velocity, specific charge, and penetration rate.

FIGURE 9.7
A power/cable shovel, Cat 7495. (Courtesy of Caterpillar, Peoria, IL.)

TABLE 9.2

Diggability Classification According to Specific Digging Energy

Digger Capacity (yd³ – m³)	Specific Digging Energy (kWh/m³)			
	Easy	Moderate	Moderately Difficult	Difficult
10–7.6	≤0.24	0.24–0.30	0.30–0.39	≥0.39
15–11.5	≤0.21	0.21–0.28	0.28–0.35	≥0.35
20–15.3	≤0.19	0.19–0.25	0.25–0.32	≥0.32
25–19.1	≤0.19	0.16–0.22	0.22–0.29	≥0.29

Source: Ceylanoglu, A. et al., Specific digging energy as a measure of diggability, *Proceedings of the 3rd International Symposium on Mine Planning and Equipment Selection,* October 15–18, Istanbul, Turkey, 5 p, 1994.

9.3.4 Draglines

Typical view of a dragline is seen in Figure 9.8. Walking draglines have been extensively developed over the past 40 years. They are mainly used for overburden handling and in some cases for handling coal also. Although the

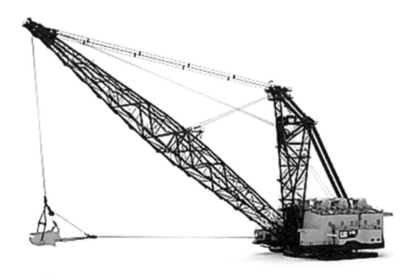

FIGURE 9.8
A typical dragline used in overburden removal in coal mines. (Courtesy of Caterpillar, Peoria, IL.)

average bucket capacity is 20 m³, the largest draglines built today utilize buckets with capacities 76 m³–115 m³. A major advantage of the walking dragline in addition to maneuverability is its extremely low bearing capacity from 60 kPa–125 kPa. Draglines are ideally suited today to the deep overburden so frequently encountered. The dragline has a lower maintenance cost than does a shovel.

Pasamehmetoglu et al. (1996) stated that in a comparison of the parameters of various draglines operating at different mine sites, the parameters such as uniaxial compressive strength of the burden to be handled, specific charge, cycle time, excavation rate, and specific energy had to be taken into account. Typical results obtained in different Turkish coal mines are given in Table 9.3. Additionally, the monitoring of bucket position was found to be important to an interpretation of the results and their relationships to the site parameters.

9.3.5 Continuous Surface Miners

Surface miners are used for selective mining of coal and/or some other minerals, thus permitting more efficient exploitation of the deposit. A rotating drum equipped with tungsten carbide cutting tools rotates in an up-milling direction and chips away the material toward the machine center from where it is transported to the discharge conveyor. Drilling and blasting

TABLE 9.3

Comparative Studies of Different Draglines Working in Turkish Coal Mines

Site	Bucket Capacity (m³)	Total Power (kW)	Compressive Strength (MPa)	Specific Charge (kg/m³)	Cycle Time (sec)	Excavation Rate (m³/h)	Specific Energy (kWh/m³)	Specific Energy per sec (kWh/m³)
1	15.30	2,498	45.78	0.166	67.63	814	0.664	0.00989
2	24.26	4,996	18.0	0.075	50.26	1,739	0.589	0.01172
3	30.60	4,996	22.63	0.143	62.48	1,744	0.602	0.00963
4	49.73	8,128	4.0	0.055	56.84	3,150	0.506	0.00890
5	53.55	8,904	13.9	0.112	66.62	2,895	0.595	0.00894

Source: Pasamehmetoglu, A.G. et al., Dragline monitoring for the determination of excavatability—A case study, *Proceedings of the Surface Mining Conference, South African Institute of Mining and Metallurgy,* September 30–October 4, pp. 195–200, 1996.

operations are no longer required, vibration, noise, and dust generation are reduced considerably. A typical surface continuous miner is illustrated in Figure 9.9 and cutting drum in Figure 9.10. Some characteristics of a widely used continuous miner are given in Table 9.4. A detailed study on coal properties effecting the selection of surface continuous miners will be given in Chapter 14.

FIGURE 9.9
A surface continuous miner working in a coal mine. (Courtesy of Wirtgen, Windhagen, Germany.)

FIGURE 9.10
Cutting drum of a surface continuous miner. (Courtesy of Wirtgen, Windhagen, Germany.)

TABLE 9.4

Some Characteristics of Wirtgen Continuous Surface Miners

Technical Properties	2100 SM	2600 SM	3500 SM	4200 SM
Total power of the machine (kW)	448	559	895	1,193
Fuel consumption at full load (1/h)	110	152	233	310
Total weight (tonne)	39.5	67.5	137	185
Width of cutting drum (m)	2	2.6	3.5	4.2
The depth of a slight (m)	0.25	0.25	0.47	0.6
Number of crawlers	4	3	4	4
Working speed (km/h)	0–1.6	0–1.5	0–1.5	0–1.5
Crawling speed (km/h)	0–4.6	0–6	0–3.9	0–2.8
Production rate up to (m³/h)	280	360	860	1,250

Source: Bilgin, N. and Balci, C., Performance prediction of mechanical excavators in underground and opencast mining, *Proceedings of the 20th World Mining Congress*, November 7–11, Tehran, Iran, pp. 75–81, 2005.

9.4 Conclusive Remarks

Workability of coal is a general name given to the cuttability, grindability, drillability, and diggability/excavability of coal. Cuttability and grindability of coal were discussed in Chapters 7 and 8. Directional drilling and diggability/excavability of coal are two other important subjects mentioned in this chapter.

In-mine directional drilling provides the coal mining industry with effective and practical options for methane drainage and exploration in advance of mining. Measurement and imaging provide for coal quality information while drilling, which increase the value of directional drilling. Bedding planes, frequency of cleats, coal permeability, and clayey inclusions within the coal seams are the major factors effecting the efficiency of in-mine directional drilling.

In recent years, unconventional gas projects oriented to CBM, shale gas, and tight gas productions showed a tremendous increase worldwide. Due to tremendous benefit gained from these projects; directional drilling, horizontal drilling, and hydraulic fracturing became one of the main subjects of the research studies directly related to the production of unconventional gas.

Surface mining operations are steadily increasing in recent years to win coal and minerals from greater depths due to the improved design of powerful mechanical excavators, such as bucket wheel excavators, hydraulic excavators, walking draglines, surface continuous miners, and stripping and loading shovels. The selection of an excavator is a prime importance because it largely determines the other equipment required and the mode of operation. The performance of these excavating or digging machines mainly depends on: (a) the design parameters of the cutting head, (b) power of the machine, (c) the intact strength and the abrasivity properties of the coal, (d) the competence of the ground as a whole, and (e) the new properties of the broken coal. It is clearly shown in this chapter that digging specific energy plays an important role in selecting any kind of excavating/digging machine.

References

Bilgin, N. and Balci, C. 2005. Performance prediction of mechanical excavators in underground and opencast mining. *Proceedings of the 20th World Mining Congress*, November 7–11, Tehran, Iran, pp. 75–81.

Brunner, D. J. and Schwoebel, J. J. 2018. Directional drilling for methane drainage and exploration in advance of mining recent advances and applications. REI Drilling, Inc., Salt Lake City, Utah, USA, 10 p. Online: http://www.reidrilling.com/dotAsset/1ef1002c-412d-40c2-a228-d287d4434b17.

Ceylanoglu, A., Karpuz, C., and Pasamehmetoglu, A. G. 1994. Specific digging energy as a measure of diggability. *Proceedings of the 3rd Int. Symp. on Mine Planning and Equipment Selection*, October 15–18, Istanbul, Turkey, 5 p.

Fuginski, Z. 2012. Underground coal mining – global picture and brief overview. *Released by Colombia Clean Power, Sas*, 14 p.

Hamrikova, R. and Jurman, J. 2011. Effective energy utilization with the excavator wheel. *Management Systems in Production Engineering*, 2(2):8–11.

https://geology.com/energy/shale-gas/, taken on 2018, what is shale gas? Republished from a December, 2010 "Energy in Brief" by the Energy Information Administration.

https://www.iea.org/ugforum/ugd/, taken on 2018, Unconventional gas production database, IEA, International Energy Agency.

Kennedy, R. E. 1990. *Surface Mining*. Society for Mining Metallurgy and Exploration, Littleton, CO, 1193 p.

Pasamehmetoglu, A. G., Bozdag, T., Ozdogan, M., Karpuz, C., and Hindistan, M. A. 1996. Dragline monitoring for the determination of excavatability: A case study. *Proceedings of the Surface Mining Conference, South African Institute of Mining and Metallurgy*, September 30–October 4, pp. 195–200.

Rasper, L. 1975. *The Bucket Wheel Excavator, Application, Development, Design*. Tran Tech publications, Germany, 326 p.

Singh, J. 1993. *Heavy Construction, Planning Equipment and Methods*. Balkema, Rotterdam, the Netherlands, 1084 p.

Sladkova, D. 2011. The utilization of GNSS technology for mining. *Process Automation, Management Systems in Production Engineering*, 2(2):3–7.

Stefanko, R. 1983. *Coal Mining Technology, Theory and Practice*. Society of Mining Engineers of the American Institute of Mining, Metallurgical, and Petroleum Engineers, Littleton, CO, 410 p.

Thomson, S. 1997. Directional drilling for safety in coal mining. In: Doyle, Moloney, Rogis & Sheldon, editors. *Proceedings of the Symposium on Safety in Mines: The Role of Geology,* November 24–25, pp. 77–84.

Yong, Q. and Jianping, Y. 2015. A review on development of CBM industry in China. *Presentation given at: AAPG Asia Pacific Region, Geoscience Technology Workshop, Opportunities and Advancements in Coal Bed Methane in the Asia Pacific,* Brisbane, Queensland, Australia, February 12, 31 p.

Wang, F., Ren, T. X., Hungerford, F., Tu, S., and Aziz, N. 2011. Advanced directional drilling technology for gas drainage and exploration in Australian coal mines. *First International Symposium on Mine Safety Science and Engineering, Procedia Engineering,* 26:25–36.

10

Effect of Coal Cutting Methods on Final Properties of the Produced Coal that would be Relevant to Later Coal Upgrading and Utilization

10.1 Introduction

During the formation of coal, some bands of clay and rock or igneous intrusions may be acquired by coal seams. In addition, during the mining operations, a part of the roof and floor material may be excavated with the coal seam also, in order to create adequate working height for the miners and equipment. Therefore, run-of-mine (ROM) coal has impurities associated with it. On the other hand, the market may demand predetermined specifications depending on the intended use of the coal such as a certain size distribution, less sulfur content, and high calorific value leading a need of less impurities etc. Certainly, a coal preparation process is needed for demanded specifications of a customer.

The size distribution of the coal obtained from mechanized face effect directly coal preparation parameters such as variation in crushing size, relative density of froth flotation, and equipment required to obtain maximum washability of coal. Coal preparation processes are dependent on the differences in physical properties between the coal components or macerals. The major differences between the lithotypes, which are dependent on their macerals, are in toughness, relative density, and crushability. Vitrain and fusain are more brittle and lighter in density than that of other lithotypes. The effect of petrographic composition on coal breakage is evident in preparation plant as well as in the mine. Cutting of coal is automatically a natural separation and results in selective concentration of certain macerals, group macerals, and lithotypes. Significant petrographic variations may therefore be recognized by screening, sizing, oil flotation, and relative density separations. In other words, the vitrain, which is best coking component, has low density, clarain and clarodurite, is best for coal conversion and the heavy fraction (inertinite and durain) is best for power plants (Falcon 1978, Falcon and Falcon 1987a, 1987b).

In the light of the statements given above, to clarify the subject in detail, Imbat coal mine located in the Soma District of Turkey, has been selected as a pilot mine, where shearers are used for coal (lignite) production. Physical, mechanical, and other properties of the coal in this mine are given in Chapters 5 and 6 and mining method in Chapter 13. Systematic sampling of the excavated coal was realized from the chain conveyor close to the shearer. Samples were sieved and petrographic analysis was carried out for certain sieve sizes.

10.2 Effect of Coal Cutting on the Size Distribution of Run of Mine Coal

Optimizing the fines content of the ROM coal offers numerous savings and benefits. The cost of washing coal fines is higher because of the processes used and the product losses that occur (resulting—as well—in a lower recovery). With the increased losses, more tailings have to be disposed in a suitable facility. A lower level of fines in the ROM results in a lower level of respirable and airborne dust, increasing workplace safety and reducing the risk of coal dust explosions. It is of the utmost importance to reduce the level of fines to the greatest possible extent. It is important to optimize the mining activity for target size material, enabling most of the material to be processed in the cheaper coarse circuit of the processing plant. It is also important to reduce crushing costs.

10.2.1 Experience Gained by Wirtgen Company

In recent years, Wirtgen Company has conducted several large-scale field tests on particle size distribution and material degradation for ROM coal, mined with continuous surface miners. Coal and sedimentary ore from eight different pits were analyzed to obtain their particle size distribution. Almost 8,000 tons of material was screened to compare the material produced by surface miners and conventional mining methods. The results are plotted in Figure 10.1, showing the advantage of surface continuous miner compared to dozer rip and push mining method. As seen in Figure 10.1, it is proved that the continuous surface miner delivers more than 70% target size coal (2 mm–40 mm), while the dozer ranges at less than 58%. Additionally, savings are generated in the crushing stage: only 17% of the material coming from the surface miner has to be crushed, as opposed to more than 26% when processing dozer coal. This will also result in more fines that have to be processed. High levels of coal fines in the ROM material result in higher costs for washing, lower recovery, reduced workplace safety, and negative impacts

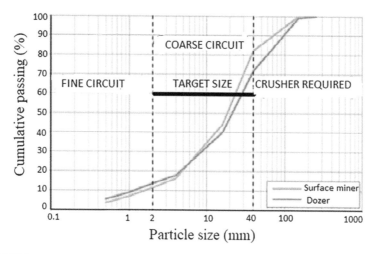

FIGURE 10.1

Particle size distribution curve obtained for Wirtgen continuous surface miners, https://www.wirtgen-group.com/en/news-media/press-releases/wirtgen-surface-mining-maximizing-coal-recovery-by-minimizing-fines.106243.php, obtained in June 2018.

on the downstream processes (https://www.wirtgen-group.com/en/news-media/press-releases/wirtgen-surface-mining-maximizing-coal-recovery-by-minimizing-fines.106243.php, obtained in June 2018).

10.2.2 Experience Gained in Imbat Mechanized Lignite Mine, Turkey

Systematic coal specimens were collected from the chain conveyor just in front of the shearer, dried, and sieved in the laboratory, the results are given in Table 10.1 and in Figure 10.2. It is interesting to note that Figures 10.1

TABLE 10.1

Sieve Analysis Obtained on Run of Mine Coal from Imbat Lignite Mine

No	Sieve Size (mm)	Weight of Coal (g)	Percent Retained %	Cumulative Weight Retained (g)	Cumulative Percent Retained %
x1	+25	3292.5	26.4	3292.5	26.4
x2	−25 +8	3371.0	27.1	6663.5	53.5
x3	−8 +2	3234.5	26.0	9898.0	79.4
x4	−2 +0.5	1687.0	13.5	11585.0	93.0
x5	−0.5 +0.125	689.5	5.5	12274.5	98.5
x6	−0.125	188.0	1.5	12462.5	100.0

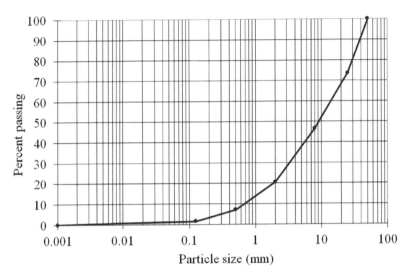

FIGURE 10.2

Particle size distribution curve obtained on run of mine coal from Imbat mine.

and 10.2 show the advantage of cutting coal mechanically having a similarity in a way that the 70%–80% of the sieved material falls within the range of 2 mm–40 mm, which is determined a target size by Wirtgen Company to minimize the cost of coal preparation plants. These results are typical example on the benefit of using mechanical coal excavation for natural sorting of ROM coal to meet the needs of coal market.

Detailed chemical analyses were carried on the coal samples for each size group sieved, the results are given in Table 10.2.

The variations of ash content and upper calorific values for different particle sizes are seen in Figure 10.3. As seen from this figure, the ash content

TABLE 10.2

Chemical Analysis Results and Calorific Values of the Coal Samples for Different Particle Sizes

Sieve Size (mm)	Humidity (%)	Volatile Matter (%)	Ash Content (%)	Dependent Carbon (%)	Upper Calorific Value (kcal/kg)	Carbon (%)	Total Sulfur (%)
+25	9.1	36.0	13.0	42.0	5648	58.8	2.0
−8 +2	9.1	33.4	22.4	35.1	4859	53.9	1.4
−2 +0.5	7.4	32.8	27.0	32.9	4416	52.9	1.4
−0.125	5.6	27.5	45.2	21.8	3244	39.0	1.1

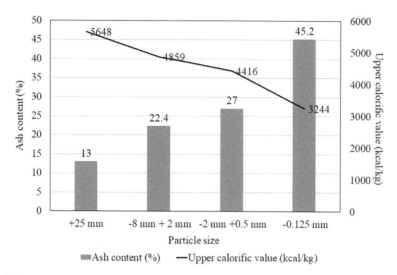

FIGURE 10.3
Variation of ash content and upper calorific value of coal obtained for different particle sizes.

of coal increases with decreasing size, and the relation is vice versa for the caloric values of coal.

The variation between volatile matter and carbon content for different particle sizes is plotted in Figure 10.4. For target size of ROM coal, the volatile matter and the carbon content have the highest values.

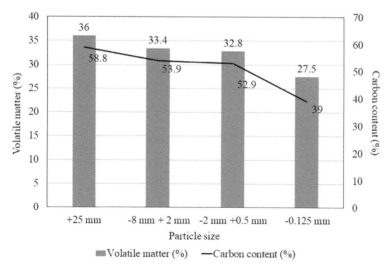

FIGURE 10.4
Variation of volatile matter and carbon content with particle size.

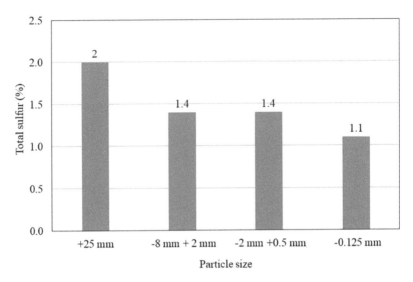

FIGURE 10.5
Variation of total sulfur with particle size.

One of the most important relations is obtained between sulfur content and particle size. As seen in Figure 10.5, the sulfur content of coal decreases considerably with decreasing particle size.

From rock or coal mechanics point of view, it is evident that the size of cut material may be adjusted by changing depth of cut, for example, pulling force of shearers showing the possibility of optimizing the need of a customer (Bilgin et al. 2014). However, it is important to note that the relations given in Figures 10.3 through 10.5 cannot be generalized, since such relations depend on the structural and petrographic characteristic of the coal.

10.3 Petrographic Characteristics of the Coal Samples

One of the main objectives of writing this chapter was to emphasize on the importance of petrographic properties of coal on the particle size distribution of ROM coal subjected to mechanical cutting in the mine, as explained in the section "10.1 Introduction."

FIGURE 10.6
Samples subjected to petrographic analysis.

Samples of four size fractions, +25 mm, −8 +2 mm, −2 +0.5 mm, and −0.125 mm, as seen in Figure 10.6, were subjected to petrographic analysis. Macerals showed typical characteristics of huminites, which is a term used for maceral group of lignite (brown coals). To clarify the subject, the subdivision of the maceral group huminite, after Sykorava et al. (2005), is given in Table 10.3. The results of petrographic analysis are given in Table 10.4.

TABLE 10.3

Subdivision of the Maceral Group Huminite

Maceral Group	Maceral Subgroup	Maceral	Maceral type
Huminite	Telohuminite	Textinite	
		Ulminite	
	Detrohuminite	Attrinite	
		Densinite	
	Gelohuminite	Corpohuminite	Phlobaphinite
			Pseudophlobaphinite
		Gelinite	Levigelinite
			Porigelinite

Source: Sykorova, I. et al., *Int. J. Coal Geol.*, 62, 85–106, 2005.

TABLE 10.4

Variation of Macerals with Sieve Size for Coals from Imbat Mine

Maceral	Particle Size (mm)			
	+25	−8 +2	−2 +0.5	−0.125
Textinite	16.2	12.7	10.1	6.3
Ulminite	44.1	35.9	34.6	32.6
Telohuminite	*60.3*	*48.6*	*44.7*	*38.9*
Attrinite	5.5	7.2	7.7	13.1
Densinite	7.9	8.1	6.1	3.5
Detrohuminite	*13.4*	*15.3*	*13.8*	*16.6*
Levigelinite	–	–	–	–
Porigelinite	1.5	1.0	1.0	0.5
Corpohuminite	2.9	3.3	3.4	2.7
Gelohuminite	*4.4*	*4.3*	*4.4*	*3.2*
Huminit	*78.1*	*68.2*	*62.9*	*58.7*
Fusinit	–	0.1	–	1.0
Semifüsinite	–	–	–	–
Inertodetrinite	0.3	1.4	1.1	2.3
Macrinite	–	0.2	0.2	–
Funginite	0.1	–	–	–
Inertinite	*0.4*	*1.7*	*1.3*	*3.3*
Cutinite	0.1	0.1	–	–
Resinite	1.9	1.4	1.3	1.0
Liptodetrinite	2.4	1.4	1.5	2.4
Sporinite	1.4	0.5	0.5	0.6
Liptinite	*5.8*	*3.4*	*3.3*	*4.0*
Minerals	*15.7*	*26.7*	*32.5*	*34.0*

As seen from this table, the percentage of huminites decreases from 78.1% to 58.7% for sieve fractions from +25 mm to −0.125 mm.

The degree of humification and especially gelification of huminite in coal affects most of the industrial processes such as briquetting, carbonization, liquefaction, gasification, and combustion. Humification is a process of formation of humic substances (organic matter that has reached maturity) decomposed from plant remains. Gelification succeeds humification at depth in which the humic substances pass through a soft, plastic stage from peat to brown coal (lignite).

Table 10.4 shows that the percentage of ulminites decreases from 44.1% to 32.6% for sieve fractions from +25 mm to −0.125 mm. The technological properties of ulminite depend on its degree of gelification. Since gelification increases the hardness of huminite macerals and drying promotes the formation of fissures, ulminite has a better grindability than textinite. High content of ulminite prevents development of links between coal grains during binderless briquetting. The briquettes have a low strength. Ulminite produces lower yields of tar and gas and higher amounts of char than textinite does. The reflectance of ulminite is a reliable rank. It correlates well with other rank-sensitive parameters such as calorific value or carbon content (Sykorava et al. 2005).

As seen from Table 10.4, textinite decreases from 16.2% to 6.3% for sieve fractions from +25 mm to −0.125 mm. Textinite influences the technical properties of lignites markedly only where presents in large amounts. With regard to its elasticity, textinite rich coal is difficult to grind. Ungelified wood composed of textinite breaks into fibrous pieces making the sieving difficult. Therefore, textinite is concentrated in the coarse size fractions. Textinite increases the briquetting properties of low rank lignites with moisture contents. In general, briquettes produced from low rank lignites containing textinite are of high strength. Due to its high content of cellulose and/or resin, textinite produces high yields of tar. The char yield is rank-dependent. Textinite-rich low-rank lignites with poor milling properties can create problems. Depending on the amount of elongated fibrous grains, the nozzles may be blocked. The combustion of such big grains is incomplete. Textinite is relatively resistant against weathering. The texture of textinite can be used for identification of certain plants and therefore for stratigraphic correlations (Sykorava et al. 2005).

Figures 10.7 and 10.8 show the views of macerals under microscope.

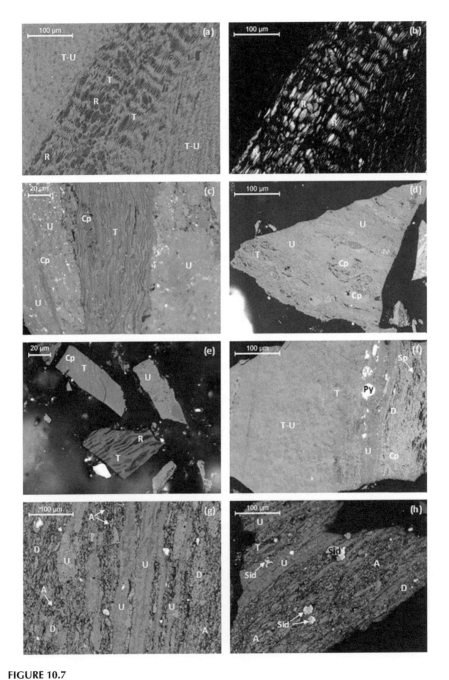

FIGURE 10.7
View of some macerals under normal light (a, c, d, e, g, and h) and under fluorescent light (b), textinite (T), textoulminite (T-U), ulminite (U), densinite (D), attrinite (A), corpohuminite (Cp), resinite (R), sporinite (Sp), pyrite (Py), and siderite (Sid).

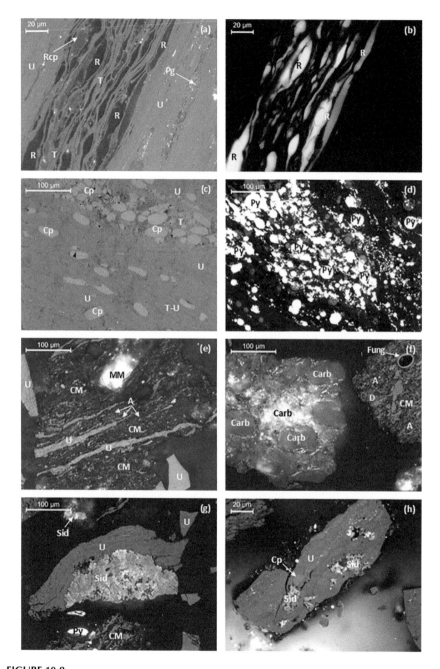

FIGURE 10.8
View of some macerals under normal light (a, c, d, e, g, and h) and under fluoeresant light (b), textinite (T), textoulminite (T-U), ulminite (U), densinite (D), attrinite (A), porigelinite (Pg), corpohuminite (Cp), funginite (fung), resinite (R), pyrite (Py), clay minerals (CM), carbonate (Carb), siderite (Sid), and minerals (MM).

10.4 Conclusive Remarks

This chapter showed that the particle size of run-of-mine coal obtained from mechanical excavation is ideal for coal preparation process. For a typical example, ash content of coal increased with particle size and the relation is vice versa for caloric values of coal. One of the most important relations is obtained between sulfur content and particle size, sulfur content of coal was found to decrease considerably with particle size.

From rock or coal mechanics point of view, it is evident that the size of cut material may be adjusted by changing depth of cut, for example, pulling force of shearers showing the possibility of optimizing the need of a customer.

A typical example obtained for a Turkish lignite mine showed that the percentage of huminites decreased from 78.1% to 58.7% for sieve fractions from +25 mm to −0.125 mm. The degree of humification and especially gelification of huminite in coal affects most of the industrial processes such as briquetting, carbonization, liquefaction, gasification, and combustion. Petrographic properties are proved to be relevant to later coal upgrading and utilization.

References

Bilgin, N., Copur, H., and Balci, C. 2014. *Mechanical Excavation in Mining and Civil Industries*. CRC Press/Taylor & Francis, Boca Raton, FL.

Falcon, L. M. and Falcon, R. M. S. 1987. The petrographic composition of Southern African coals in relation to friability, hardness and abrasive indices. *Journal of South African Institute of Mining and Metallurgy* 87(10):323–336.

Falcon, R. M. S. 1978a. Coal in South Africa, part II: The application of coal petrography to the characterization of coal. *Minerals Science Engineering* 10(1):28–52.

Falcon, R. M. S. 1978b. Coal in South Africa, part III: The fundamental approach to the characterization and rationalization of South Africa Coal. *Minerals Science Engineering* 10(2):130–153.

https://www.wirtgen-group.com/en/news-media/press-releases/wirtgen-surface-mining-maximizing-coal-recovery-by-minimizing-fines.106243.php.

Sykorova, I., Pickel, W., Christanis, K., Wolf, M., Taylor, G. H., and Flores, D. 2005. Classification of huminite—ICCP system 1994. *International Journal of Coal Geology*, 62:85–106.

11

Mechanized Underground Coal Exploitation Methods: Longwall, Room and Pillar and Shortwall

11.1 Introduction

Geology, quality of the coal seams, and depth are the factors that dictate which of the surface or underground methods is the most cost efficient method of coal exploitation. In addition to factors such as density of the overburden or thickness of the seam, depth of the deposit is of vital importance for selection between underground and surface mining. Except for some infrequent cases, the deposits deeper than 50 m–100 m are usually mined by underground methods. The criteria such as what follows are of use for technical or economic feasibility assessment of a coal mining project: regional geological conditions, properties of the overburden and the coal seam, ground water conditions, capital costs, topography, strength of the material above and below the seam, climate, availability of labor and materials, customer requirements, and effective property control over the area to be mined, as it has an effect on availability of land for mining and access.

Modern mechanized underground coal exploitation methods can be classified into three categories as room and pillar, longwall, and shortwall. Since its birth, longwall mining has progressively gained significance as one of the chief coal mining methods. While, in comparison with the other underground mining techniques, its production rate is currently greater, the gap tends to widen as technology makes progress. Although once recognized as the conventional underground coal mining method, room and pillar mining has become accustomed to modern machinery and is currently the method of choice for mining sedimentary deposits. Shortwall mining, although similar to longwall mining in the sense that both methods use mechanical miners with moveable roof support, is responsible for less than 1% of deep coal production.

Almost half of the world underground coal production comes from longwall mining. The rest of the coal production comes from room and pillar, shortwall, conventional, and other methods. In 2016, the average underground coal production of the United States was 228×10^6 tons, of which 130×10^6 tons came from longwall mining method. The production efficiency was 4.6 tons per employee for longwall mining, 2.45 tons per employee for room and pillar continuous mining, and 9.3 tons per employee for conventional and other methods (shortwall, scoop loading, hand loading, etc.) (EIA 2017).

11.2 Longwall Mining Method

Longwall mining technology has been established as an underground coal exploitation method worldwide since its inception in the seventeenth century in the United Kingdom (Das 1999). The modernized form of longwall mining was established at 1963 (Galvin 2016). Beginning from 1980s, the method has constantly evolved such that it has become the most dependable and efficient method of underground coal mining for deep deposits, which are economically not mineable using surface mining methods. Compared to room and pillar method, longwall mining has the potential to reach the significantly higher recovery rate of 75%. The machinery cost, however, is more than that of room and pillar method. The primary advantages and disadvantages of longwall mining are summarized in Table 11.1.

TABLE 11.1

Advantages and Disadvantages of Longwall Mining Method

Advantages	Disadvantages
Low operating cost	High capital investment
High productivity	Significant advance development
High recovery	Low selectivity
Safest method	Predictable subsidence
High production rate	Low flexibility
High mechanization	
Continuous method	
Short training period	

Source: Kumar, A. et al., The systems/ground control interface and its impact on mining method selection, in: *Underground Mining Methods and Technology*, Szwilski, A. B. and Richards, M. J. (Eds), Elsevier Science Publishers, B.V., Amsterdam, the Netherlands, 321–332, 1987; Nieto, A., Soft-rock (underground) mining: Selection methods, *SME Mining Engineering Handbook*, 3rd edition, Chapter 29(7.4), 389–396, 2011.

Longwall mining is a commonly used method of coal extraction that involves the complete removal of large, rectangular panels of coal (EIA 1995). Longwall mining is divided into two major groups: "longwall mining on the advance" and "longwall mining on the retreat." The first requires constructing maingate and tailgate access in front of the face. The second, which is the most popular type, requires establishment of two to three gate roads that are parallel to the mining direction and meet against the face (Galvin 2016).

In modern longwall mining method, the face and panel lengths change between 100 m and 300 m and 600 m and 2,500 m, respectively. The longwall mining has three main components: an excavation machine, hydraulic roof supports, and armored face conveyor (AFC). Excavating machines used in this method are plows and shearers that move back and forth across the coal seam. AFC is used in order to remove the excavated coal from the work area. The coal removed falls onto AFCs, which also support the weight of the excavating machines. AFC carries coal to a belt conveyor in order to move it out of the mine. In the longwall mining, the roof is supported by hydraulic roof supports for overlying the rock. Hydraulic roof supports move forward as the extraction progresses along the panel. While moving the longwall mining equipment forward, the rock and roof are generally allowed to fall/collapse behind the operation. The safety and production in longwall mining increased with this supporting system (Bilgin et al. 2014). A schematic view of longwall mining on the retreat system is illustrated in Figure 11.1.

While there are many criteria to consider in deciding whether to select a shearer or a plow as excavation equipment, seam thickness is probably one of the most important (Stefanko 1983). The vast majority of modern longwalls utilize double-drum shearers. There is no doubt choosing double-drum shearers in thick seams. However, single-drum shearers can be used for seams less than 1 m thickness. Plows (ploughs) are especially utilized in very thin seams.

A shearer usually consists of the following main components: shearer drum, ranging arm, self-haulage system, cowls, bretby cable handler, and control system. Shearers can be used in weak and hard coal seams with the heights from 1.5 m up to 7.0 m, and they may operate in faces with dip angles up to 20°. Their height can easily be adapted to changeable seam height. The installed power of the modern shearers is up to 2,800 kW. They can excavate coal with haulage speeds of up to 41 m/min and excavate with a drum width (web) between 0.8 m and 1.2 m. Specific energy consumed in cutting changes between 0.7 MJ/m^3 and 10.0 MJ/m^3 that changes with coal hardness (Myszkowski and Paschedag 2008). Machine utilization time of shearers is higher than that of plows and usually changes between 40% and 60%. The shearer excavates the coal in two different directional modes: bidirectional (in both directions) and unidirectional (in one direction) (Bilgin et al. 2014). Technical properties of some shearers are summarized in Table 11.2. Sample photographs of shearers are given in Figure 11.2.

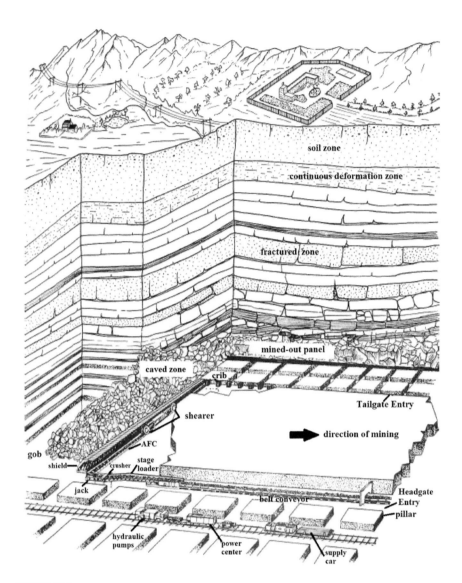

FIGURE 11.1
Schematic view of longwall mining on the retreat system. (Modified after Mishra, B., *Modern American Coal Mining: Methods and Applications*, Section 9: Longwall Mining. C. J. Bise (Ed), Published by the Society for Mining, Metallurgy, and Exploration, Inc., Ebook 978-0-87335-395-3, 2013.)

TABLE 11.2

Technical Properties of Some Shearers (Product Catalogues)

Shearers	Eickhoff		CAT	Joy	T Machinery	
	SL300	SL1000	EL3000	7LS5	MB850E	MB1200E
Seam thickness (m)	1.6–4.0	3.0–8.6	2.5–5.5	2.0–4.5	1.6–4.5	2.0–5.5
Drum diameter (m)	1.4–4.0	–	2.5–2.75	1.55–1.75	1.3–2.25	1.8–2.75
Drum width (m)	0.8	–	1.1	0.96–1.04	0.8–1.0	0.8–1.0
Rotational speed (rpm)	46–56	29–41	35–43	47–66	29–47	26–40
Haulage speed (m/min)	0–40	0–41	0–31.1	0–30	0–20	0–24
Haulage force (kN)	–	–	1,075	800	800	900
Cutting power (kW)	–	–	2 × 860	600–750	2 × 350	2 × 500
Haulage power (kW)	1–80	–	2 × 150	2 × 110	2 × 60	2 × 85
Total power (kW)	1,158	2,800	2,295	–	1,310	1,584
Weight (ton)	40–55	110–255	105	81.7	80	105

FIGURE 11.2
Eickhoff SL300 shearers with and without water spray. (Courtesy of Eickhoff.)

Plows are the other excavating machines used in longwall mining. They are mechanically moving devices with cutting tools, which shear off coal and push it onto the face conveyor. They do not have rotating parts. When compared to shearers, these machines are cheaper, simpler, and relatively generate lower dust. On the other hand, they have some disadvantages compared to shearers such as: cutting height is fixed, ability to cut is limited, machine stability becomes more problematic with increasing cutting height, etc. They have an ability to cut the seam in both directions. They can be used in coal strength of up to 40 MPa with seam height approximately less than 2.3 m, and they may operate in faces with dip angles up to 45°. When the height of the seam is less than 1.0 m, base plate plows can be used. If the thickness of the seam is bigger than 1.0 m, gliding plows can be used. In low strength coal seams, the thickness of coal seam should be bigger than the height of the plow. Plows would cut harder rocks based on the technological improvements. The installed cutting power of the modern plows is up to 2 × 800 kW with plow speed of up to 3.6 m/s. Specific energy consumed in cutting changes between 1.0 MJ/m³ and 10.0 MJ/m³. Machine utilization time of the plows is less than that of shearers under comparable conditions. Maximum cutting depths of plows per pass on the face are up to 250 mm that changes with coal hardness (Bilgin et al. 2014, Myszkowski and Paschedag 2008). Technical properties of some plows are summarized in Table 11.3. A sample photograph of plow is given in Figure 11.3.

TABLE 11.3

Technical Properties of Some Coal Plows (Product Catalogues)

Plows	CAT GH800 Plow	CAT GH1600 Plow	CAT RHH800 Plow	Zhang Akou
Seam thickness (m)	0.9–2.0	1.0–2.3	0.6–1.6	0.8–1.8
Maximum seam inclination (°)	60	60	60	25
Maximum power (kW)	2 × 400	2 × 800	2 × 400	–
Maximum plow speed (m/s)	3.0	3.6	2.5	1.7
Maximum cutting depth (mm)	150	250	150	120
Weight (ton)	–	–	–	92–225

FIGURE 11.3
CAT GH800 coal plow. (Courtesy of Caterpillar, Peoria, IL.)

11.3 Room and Pillar Mining Method

Room and pillar mining has been practiced in the United States since the late 1700s, and it continues to be used profitably today. By using a continuous underground miner, this method almost extracts 50% of a coal deposit. However, it leaves enough coal in the pillars in order to support weight of the overlying strata (Schmid & Company, Inc. 2016). Since at greater depths larger pillars that reduce the coal recovery are needed, room and pillar mining generally is limited to depths of about 300 m (EIA 1995). The primary advantages and disadvantages of room and pillar mining are summarized in Table 11.4.

In room and pillar mining, several parallel entries are driven into the coal seam. These entries are then connected at intervals by wider entries, called rooms. The mine is split into a series of 6 m–10 m rooms with pillars up to 30 m wide. Production per shift is usually between 300 tons and 500 tons (Bilgin et al. 2014). A schematic view of room and pillar mining is seen in Figure 11.4.

Conventional mining or continuous underground miners are used to extract the coal in this method. The "continuous" version of this method is the most common, a continuous underground miner excavates the coal and loads it into shuttle cars or onto conveyors. Continuous mining makes cycling of equipment and its associated delays less of a problem by combining the unit operations of drilling and blasting, cutting, and loading into one operation by a single machine (Stefanko 1983). Continuous underground miners account for about 45% of the underground coal production in the room and pillar mining method (EIA 1995). They can be used in weak and hard coal seams

TABLE 11.4

Advantages and Disadvantages of Room and Pillar Mining Method

Advantages	Disadvantages
Continuous production	Moderate capital costs
Rapid development rate	Limitation on depth
Excellent ventilation	Moderate selectivity
High productivity	Variable subsidence
Moderate operating cost	Higher cost with partial extraction
Good recovery (with pillar extraction)	Moderate recovery (without pillar extraction)
Flexible	Longer training period
	Gradient limits

Source: Kumar, A. et al., The systems/ground control interface and its impact on mining method selection, In: *Underground Mining Methods and Technology*, Szwilski, A. B. and Richards, M. J. (Eds), Elsevier Science Publishers, B.V., Amsterdam, the Netherlands, 321–332, 1987; Nieto, A., Soft-rock (underground) mining: Selection methods, *SME Mining Engineering Handbook*, 3rd edition, Chapter 29(7.4), 389–396, 2011.

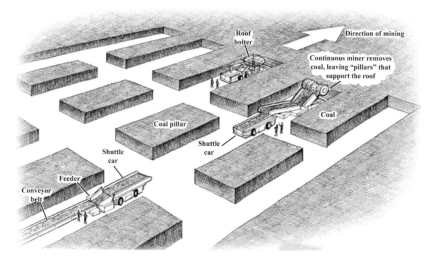

FIGURE 11.4
Schematic view of room and pillar mining method. (Modified after Arch Coal, Annual report pursuant to section 13 or 15(d) of the Securities Exchange Act of 1934, Retrieved from: https://www.sec.gov/Archives/edgar/data/1037676/000104746912001909/a2207536z10-k.htm, 2009.)

with seam heights from 0.8 m up to 6.0 m. The installed power of continuous miners is up to 950 kW. Remote-controlled continuous underground miners are used in a variety of difficult seams and conditions and robotic versions controlled by computers are becoming increasingly common (Bilgin et al. 2014). Technical properties of some continuous miners are summarized in Table 11.5. A sample photograph of continuous miner is given in Figure 11.5.

TABLE 11.5

Technical Properties of Some Continuous Miners

Continuous miners	CAT	Joy		Sandvik		Eickhoff	
	CM445	12HM37	14HM27	MC250	MC470	CM2H30	CM2H45
Drum diameter (m)	1.12	1.32	1.12	0.95	1.4	1.2	1.2
Drum width (m)	3.58	3.6	3.8	2.7	3.8	3.5	3.5
Weight (ton)	85.7	132	72	44	126	68	78
Maximum cutting height (m)	3.79	5.1	3.4	3.6	5.0	3.0	4.5
Minimum cutting height (m)	1.5	2.9	1.3	1.8	2.8	1.4	2.2
Cutting power (kW)	2 × 209	2 × 260	2 × 220	132	2 × 270	2 × 150	2 × 180
Total power (kW)	727	889	705	267	930	560	620
Rotational speed (rpm)	52	42	50	–	–	48	48

FIGURE 11.5
CAT CM445 continuous miner. (Courtesy of Caterpillar, Peoria, IL.)

11.4 Shortwall Mining Method

Shortwall mining was introduced into Australia in 1959 (Galvin 2016) to take advantage of the recent development of suitable hydraulic longwall supports, coupled with the productivity and low capital cost of continuous miners and shuttle cars. Shortwall mining represents a compromise between the longwall and room and pillar systems. Technical, economical, and social factors must be taken into consideration in deciding whether or not to use the shortwall (Stefanko 1983). Shortwall mining, occasionally used while the mine is being converted to a full longwall mine, is meant to improve the performance of continuous miners with a rather low capital cost. On occasions, because of its relatively higher adaptability, shortwall mining replaces longwall mining when it is impossible to use due to either an inconsistent coal seam or adverse mine geometry. The layout of shortwall mining is almost the same as that of longwall mining, with the exception that in shortwall mining the faces are substantially shorter than 150 m. In this method, similar to longwall mining, an installation artery is first excavated, followed by support installation. Then, the face is cut in increments of drum widths by continuous underground miners, while the extracted coal is hauled from the face to the belt conveyor by shuttle cars rather than an AFC. The primary advantages and disadvantages of shortwall mining are summarized in Table 11.6. A layout of shortwall mining is seen in Figure 11.6.

TABLE 11.6

Advantages and Disadvantages of Shortwall Mining Method

Advantages	Disadvantages
Self-advancing hydraulic roof support	Capital intensive
Good productivity	Inflexible
Roof strong enough to support wide exposed area	Maximum surface subsidence
Excellent ventilation	High operating cost
Increased safety	Medium mechanized
Short training period	Medium thickness of seam
	Soft coal only with hard floor

Source: Singh, R.D., *Principles and Practices of Modern Coal Mining*, New Age International, 696 p, 1997; Kumar, A. et al., The systems/ground control interface and its impact on mining method selection, In: *Underground Mining Methods and Technology*, Szwilski, A. B. and Richards, M. J. (Eds), Elsevier Science Publishers, B.V., Amsterdam, the Netherlands, 321–332, 1987.

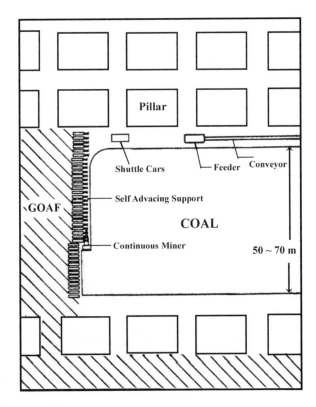

FIGURE 11.6
Layout of a shortwall panel. (Modified after Stefanko, R., *Coal Mining Technology: Theory and Practice.* American Institute of Mining, Metallurgical, and Petroleum Engineers, New York, 1983.)

11.5 Conclusive Remarks

Automated mechanized production is considered as the basic solution to supply the required high coal demand and reduce safety problems. Technological developments provided eventually for highly mechanized underground coal mining methods. These methods have been gradually focused on three basic methods: longwall, room and pillar, and shortwall mining methods. Among them, the longwall mining method provides basically highest productivity with lowest safety problems. However, it should be noted that the level of technology and automation require further developments to reach the final target of manless production systems.

References

Arch Coal, Inc. 2009. Annual report pursuant to section 13 or 15 (d) of the Securities Exchange Act of 1934. Retrieved from: https://www.sec.gov/Archives/edgar/data/1037676/000104746912001909/a2207536z10-k.htm.

Bilgin, N., Copur, H., and Balci, C. 2014. *Mechanical Excavation in Mining and Civil Industries*. CRC Press/Taylor & Francis Group, Boca Raton, FL.

Das, S. K. 1999. *Bord and Pillar Mechanism in India*. IIT Karagpur, India.

Energy Information Administration (EIA). 1995. Longwall mining briefing note. Report to the Office of Coal, Nuclear, Electric and Alternate Fuels, US Department of Energy, March 1995, DOE/EIA TR 0588.

Energy Information Administration (EIA). 2017. Annual coal report 2016. Report to the Office of Independent Statistics & Analysis, US Department of Energy, November 2017.

Galvin, J. M. 2016. *Ground Engineering: Principles and Practices for Underground Coal Mining*. Springer, Cham, Switzerland, 684 p.

Kumar, A., Haycocks, C., and Unal, A. 1987. The systems/ground control interface and its impact on mining method selection. In: *Underground Mining Methods and Technology*. Szwilski, A. B. and Richards, M. J. Eds. Elsevier Science Publishers, B.V., Amsterdam, the Nertherlands, 321–332.

Mishra, B. 2013. *Modern American Coal Mining: Methods and Applications*. Section 9: Longwall Mining. C. J. Bise (Ed), Published by the Society for Mining, Metallurgy, and Exploration, Inc. Ebook 978-0-87335-395-3.

Myszkowski, M. and Paschedag, U. 2008. *Longwall Mining in Seams of Medium Thickness, Comparison of Plow and Shearer Performance under Comparable Conditions*. Bucyrus, Ohio, 28 p.

Nieto, A. 2011. Soft-rock (underground) mining: Selection methods. *SME Mining Engineering Handbook*, 3rd edition. Chapter 29(7.4):389–396.

Schmid & Company, Inc. 2016. Longwall mining A to Z: Learning from the Pennsylvania experience. *The Illusion of Environmental Protection: Permitting Longwall Coal Mines in Pennsylvania.* Prepared for Citizens Coal Council, Canonsburg, PA, 45 p.

Singh, R. D. 1997. *Principles and Practices of Modern Coal Mining.* New Age International (P) Ltd. 696 p.

Stefanko, R. 1983. *Coal Mining Technology: Theory and Practice.* American Institute of Mining, Metallurgical, and Petroleum Engineers, New York.

12

Cutting Coal with Plows, a Trend toward Mining Thinner Coal Seams

12.1 Introduction

According to 2017 data obtained from https://www.worldcoal.org/coal/ uses-coal/coal- electricity, coal plays a vital role in electricity generation worldwide. Today, coal fuels 37% of the world's electricity generation, and this percentage is expected to last for the next several decades. In the past, thick and moderately thick coal seams were mined intensively, and today there is a tendency to produce thinner coal seams.

Churth claimed in 1981 that, in the eastern United States, 49 billion tons (Bt) or 29% of a coal reserve base of 169 Bt minable by underground methods fall in the 0.7 m–1.1 m of thickness range. The underground reserve base includes bituminous and anthracite coalbeds more than 0.7 m in thickness to a maximum depth of 300 m, and thin coal seams will become of increasing interest in the future (Churth 1981).

Wang et al. (2012) claim that coal seams with thickness less than 2.0 m are regarded as thin seams in the United States of America and other western countries, but the international definition of thin seams is used for coal seams with thickness of 0.8 m–2.0 m range. The reserves of thin coal seams, which are around 1.3 m in thickness, are enormous in China. Among 95 national key coal enterprises, a total of more than 750 thin coal seams exist in 445 coal mines. The recoverable reserves of thin seams are approximately 6.5 Bt, which account for 19% of the total recoverable coal reserves (Chen et al. 2016).

Shearers are commonly used in seam thicknesses as low as 1.5 m and as high as or even higher than 6.0 m. In the past, there were many attempts to use shearers cost-effectively in seams lower than 1.5 m. Most of those attempts were unsuccessful in terms of efficiency and cost-effectiveness. However, plows work in seams from 0.6 m to 2.3 m effectively. Generally, seams below 1.0 m use base plate plows. In thicker seams, gliding plows are found to be more effective. Because of their lower height, plows are able to mine in-seam to extract coal without cutting bed rock (Myszkowski and Paschedag 2013). The main features differentiating cutting coal with shearers and plows are given in Table 12.1. Base plate plow for seam thickness of 80 cm–160 cm is seen in Figure 12.1.

TABLE 12.1

Important Factors Regarding the Technical Applicability of Shearers and Plows

		Shearer	Plow
1. Seam thickness		1.5 m up to 6.0 m+	0.6 m up to 2.3 m
2. Coal hardness		Both types of coal extraction systems are comparable	
3. Inclination	Face	Up to 20°	Up to 45°
	Panel	Uphill up to 20°, downhill up to 20°	Uphill up to 45°, downhill up to 20°
4. Mining through faults		Both types of coal extraction systems are comparable	
5. Undulations		Plow can negotiate seam undulations much easier than a shearer	
6. Immediate roof		Friable roof can easily lead to roof falls	Smaller cutting depth allows safe operation even under friable roof
7. Immediate floor		The applicability of both systems is comparable (shearer shields use a base lift, plow shields use a special feature called elephant step to work in soft floor)	
8. Raw coal size		Shearer produces more fine particles	Plow produces more large lumps
9. Entry dimensions		Tail drive located inside the face	Plow normally requires wider tail entry
10. Automation		Shearers not fully automated yet	Some plow systems are full automation-capable and used worldwide

Source: Myszkowski, M. and Paschedag, U., *Longwall Mining in Seams of Medium Thickness Comparison of Plow and Shearer Performance under Comparable Conditions,* Caterpillar, Peoria, IL, 51 p, 2013.

FIGURE 12.1
Base plate plow RHH 800 for seam thickness of 80 cm–160 cm. (Courtesy of Caterpillar, Peoria, IL.)

12.2 Past and Present Uses of Coal Plows
in Different Countries

12.2.1 Turkey

An attempt of using a Westfalia Lünen (1980) plow was done in ELI (Aegean lignite mine) in Darkale, which led to make a classification of coal seams based on plowability according to mechanical properties of coal (Bilgin et al. 1992, Bilgin and Phillips 1994). The characteristics of the working area were as below:

District: between levels of +256/285;

Date: 11 March to 30 September 1991;

Seam thickness: 2 m;

Shortwall length: 60 m;

Speed of the plow: 0.8 m/sec;

Panzer conveyor speed: 1.2 m/sec;

Cutting depth: 5 cm; and

Driving power: 2 × 55 kW.

During the trial, coal samples were taken at every 5 m along the face and cube specimens of 5 cm × 5 cm × 5 cm in size were tested for uniaxial compressive strength. Also, point load strength, cone indenter, Schmidt hammer (N type), and impact strength index tests were evaluated for every 5 m. As seen from Figure 12.2, there was an enormous difference between the strength of the top and bottom levels of the seam, changing from 150 kg/cm^2 to 510 kg/cm^2. The trials were abandoned after 203 days. The application of

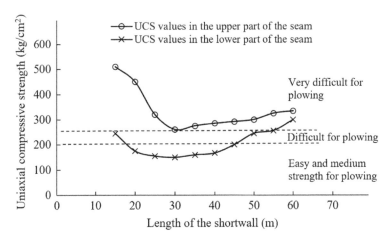

FIGURE 12.2

Variation of uniaxial compressive strength along the shortwall in Darkale. (From Bilgin, N. et al., The cuttability classification of coal seams and an example to a mechanical application in ELİ Darkale Coal Mine. *Proceedings of the 8th Coal Congress of Turkey,* Zonguldak, Turkey, 31–51, 1992; Bilgin, N. and Phillips, H. R. Mechanical properties of coal. In: *Coal, Resources, Properties Utilization, Pollution.* Editor, Orhan Kural, Istanbul, Turkey, 1994.)

plow was found inefficient in Darkale District. The inefficiency of this system was explained by the high variability of the seam strength.

This intensive study done in Darkale-Aegean led to the development of the cuttability assessments of a mechanical plow as given in Table 12.2 (Bilgin et al. 1992, Bilgin and Phillips 1994).

TABLE 12.2

Strength Classification of Coal Seams in Aegean Lignite Mine for Cuttability Assessments of a Mechanical Plow

Strength Properties	Easy to Plow	Moderate to Plow	Difficult to Plow	Very Difficult to Plow
Compressive strength (kg/cm²)	<120.0	120.0–200.0	200.0–250.0	>250.0
Point load strength (kg/cm²)	<3.5	3.5–4.5	4.6–11.0	>11.0
Cone indenter hardness	<1.5	1.5–1.8	1.8–2.6	>2.6
Schmidt hammer (N-type)	<27.0	20.0–27.0	27.0–42.0	>42.0
Impact strength index	<45.0	45.0–58.0	58.0–72.0	>72.0

Source: Bilgin, N. et al., The cuttability classification of coal seams and an example to a mechanical application in ELİ Darkale Coal Mine. *Proceedings of the 8th Coal Congress of Turkey,* Zonguldak, pp. 31–51, 1992: Bilgin, N. and Phillips, H.R., Mechanical properties of coal, In: *Coal, Resources, Properties Utilization, Pollution,* Kural, O., Ed, Istanbul, Turkey, 1994.

12.2.2 Germany

Plowing is a coal mining method invented in the early 1940s in Germany. Between the 1950s and 1980s, plows clearly dominated the German coal mining industry. In the first half of the 1990s, shearers became more capable and thus more important. Since then, shearers have outbalanced plows. This situation lasted over a decade (Myszkowski and Paschedag 2013). Many different plow models were designed and tested underground, until the mid-1990s, only two models remained, the Reisshakenhobel®, or base plate plow, and the Gleithobel®, or gliding plow. In Gleithobel plows, the plow unit is pulled by a chain on the face side of the AFC (armored face conveyor). There is no steel plate under the Gleithobel plow. Compared with the sliding plow (base-plate plow), the greater efficiency of the Gleithobel plow system allows coal to be cut by utilizing higher power. Gleithobel plows are therefore more suitable for mining coal of higher strengths and for thicker seams. Horsepower and plow speed have increased steadily over time. The PMC®-R Cat plows offer installed cutting power of up to 2×800 kW, coupled with a blazing plow speed of up to 3.6 m/s and world-leading automation capabilities (Paschedag 2014).

Typical plowing production data given in Table 12.3 were obtained from two Ruhr District Collieries, Germany in 1980s (Westfalia Lünen catalog).

TABLE 12.3

Plowing Data Obtained in Ruhr District in 1980, in Germany

Colliery	Ruhr District 1	Ruhr District 2
Seam thickness (m)	2.0–2.1	2.2–2.8
Gradient (degree)	5	14–23
Face length (m)	224	170
Shifts/day	3	3
Plow	Reisshaken, P3	Gleithobel
Driving power (kW)	$2 \times 47/120$ KW, pole changing	$2 \times 47/120$ KW, pole changing
Plow chain (mm)	24×86	26×92
Plow speed (m/sec)	2	0.43–1.3
Panzer conveyor	MIv-500	MIv-500
Driving power of conveyor (kW)	2×100	2×100
Chain (mm)	18×64 double strand chain	19×64.5
Conveyor speed (m/sec)	0.65	0.65
Saleable output (tons/day)	3,600	1,524
Rate of advance (m/day)	5.65	3.46

Source: Westfalia Lünen, Longwall mining systems, catalogue, 1980.

12.2.3 Poland

One of the most productive coal mine operating fully automated plow in thin seams is Bogdanka mine, located in Eastern Poland. It is one of the Poland's leading producers of hard coal for almost more than 30 years. Coal basin consists of nearly horizontal coal seam overlain by approximately 700 m overburden. The current mine reserve is around 255 million tons (Mt). Table 12.4 is a summary of the plow application in this mine.

12.2.4 China

Wang et al. (2012) summarized the latest developments of fully mechanized mining system with plow as below.

With geological conditions of a thin coal seam having average thickness of 1.3 m and a dip angle ranging from 5° to 8°, Xiaoqing Coal Mine of Tiefa Coal Mine Group applied the W1E-703 plow at the coal face with panel length of 905 m. Through the automatic mining system, mine produced 0.6 Mt within nine months.

Based on the geological conditions of the No. 10 coal seam with average thickness of 1.3 m, dip angle of 3°–17°, and Protodyakonov coefficient "f" of 3–4, and the immediate roof, which contains a 21 m thick sandstone layer with a Protodyakonov coefficient "f" of 8–10, Jinhuagong Coal Mine of Datong Coal

TABLE 12.4

Summary of Plow Mining Operations in Bogdanka Mine

Installed Date	March 23, 2010	October 2011
Face length (m)	250	250–305
Panel length (m)	1,750	5,022
Seam thickness (m)	1.4–1.7	1.2–1.6 (Average 1.42)
UCS (MPa)	8–19	8–19
Protodyakonov strength	0.75–1.2	0.75–1.2
Roof support	1.75 m wide 2-leg	1.75 m wide 2-leg
Installed plow power (kW)	2 × 210/630	2 × 210/630
Plow chain (mm)	42 × 137	42 × 137
Plow speed (m/s)	0.98–2.94	0.98–2.94
Plow body	Cat GH1600	Cat GH1600
Armored Face Conveyor width (mm)	1,750	1,750
AFC chain assembly (mm)	42 × 146	42 × 146
Installed AFC power (kW)	Cat AFCPF4 2 × 800	Cat AFCPF4 2 × 800
AFC speed (m/s)	1.52	1.52
Mean production (tons/day)	8,200	12,983
Peak production (tons/day)	16,894	24,900 (14.02.2012)

Source: Stopa, Z. and Myszkowski, M., Mining low coal seams, a Polish success, *World Coal*, 2015.

Mine Group employed the GH9-38Ve/5.7 plow, which was supported by the Deutsche Bergbau Technik GmbH (DBT) personnel, on a coal face with panel length of 1,050 m. Deep-hole pre-splitting blasting was applied to avoid the impact loads of large area roof weighting, with blast hole depth of 70 m and diameter of 90 mm. The coal face accomplished an output of 6,815 tons/day and 1.82 Mt per annum and was the first highly automated working face with plow in a hard thin coal seam with hard hanging wall in China.

12.2.5 USA

The number of longwall with plow mining decreased systematically in the United States of America since 1960, with lower installed power and lower degree of automation compared to today's plow capabilities. In spite of the decreasing number of plow systems in the United States of America, the predecessors of Caterpillar's plow product group (Westfalia Lünen, DBT and Bucyrus) have supplied a number of plow systems to the Pinnacle Mine near Pineville, West Virginia over the last 22 years. These plow systems have performed remarkably, even setting a world record (Myszkowski and Bauckmann 2014). Table 12.5 is a summary of plow application in this mine.

TABLE 12.5

Some Equipment Installed in Pinnacle Mine since 1989 and Mine Production

Installed Date	1989–1991	February 1999	July 2010
Face length (m)	250	319	300
Panel length (m)	—	2,800	—
Seam thickness (m)	—	1.4	1.4
Cutting depth (mm)	—	Up to 250	—
Roof support	2-leg	—	Bucyrus
Installed plow power (kW)	2 × 270	2 × 400	2 × 600
Plow chain (mm)	30 later 38	38	42
Plow speed (m/s)	1.54	1.98	1.0–1.98
Plow body	Gleithobel	Triple	—
AFC width (mm)	900	1,032	1,132
AFC chain assembly (mm)	—	—	—
Installed AFC power (kW)	2 × 270	2 × 600	2 × 600
AFC speed (m/s)	—	1.86	2.13
Mean production (ton/day)	10,889	14,000–17,000	10,000 restricted by CH_4
Peak production (ton/day)	18,145	22,700	32,400

Source: Myszkowski, M. and Bauckmann, S., *Operational Experiences with Automated Plow Systems at Pinnacle Mine in the USA,* Caterpillar Global Mining, Caterpillar, Peoria, IL, 11 p, 2014.
http://www.viewpointmining.com/article/pinnacle-mine-breaks-world-record, taken on May 2018, Viewpoint Perspectives on Modern Mining Issue 11.

The Pinnacle coal mine lies in southwestern West Virginia, near the city of Pineville. All exploitation of black coal at this mine is carried out in the Pocahontas #3 seam. This seam lies at a depth between 300 m and 500 m and has a thickness of 1.1 m–1.4 m (1.25 m on average). All mine activities are in horizontal deposits, with an inclination of less than 5°. The coal in the Pocahontas seam is classified as "easily plowable." While the face length was fixed to a maximum of 250 m, the panel length was limited to 2,200 m for firedamp prevention reasons (Myszkowski and Bauckmann 2014, http://www.viewpointmining.com/article/pinnacle-mine-breaks-world-record).

12.3 Conclusive Remarks

Plowing is a coal mining method invented in the early 1940s in Germany. Between the 1950s and 1980s, plows clearly dominated the German coal mining industry. In the first half of the 1990s, shearers became more capable and thus more important. Since then, shearers have outbalanced plows. This situation lasted over a decade. However, in the past, thick and moderately thick coal seams were mined intensively all over the world, and today there is a tendency to produce thin coal seams. Churth claimed in 1981 that in the eastern United States, 49 Bt of coal fall within the thickness range of 0.7 m–1.1 m. The recoverable reserves of thin seams in China are approximately 6.5 Bt, which accounts for 19% of the total recoverable coal reserves (Chen et al. 2016). However, plows work effectively in seams from 0.6 m to 2.3 m in thickness. Generally, seams below 1.0 m, use base plate plows. In thicker seams, gliding plows are found to be more effective (Myszkowski and Paschedag 2013).

It is reported that fully automated plowing systems in coal seams having thickness of 1.2 m–1.8 m opens new horizons in mechanized coal winning with mean daily production rates changing between 8,000 and 17,000 tons/day as in Bogdanka (Poland) and Pinnacle (United States of America) coals. As seen in Table 12.4, the compressive strength of coal seams in Bogdanka Mine changes between 8 MPa and 19 MPa or 80 kg/cm² and 190 kg/cm², which coincides well with the plowability classification developed for Darkale Mine, Turkey (Table 12.2).

As a conclusion, fully automated plow systems opened new horizons for mining thin coal seams very efficiently even breaking world records for mechanized coal mining.

References

Bilgin, N. and Phillips, H. R. 1994. Mechanical properties of coal. In: *Coal, Resources, Properties Utilization, Pollution.* Editor, Orhan Kural, Istanbul, Turkey, ISBN: 975-95701-1-4.

Bilgin, N., Phillips, H. R., and Yavuz, N. 1992. The cuttability classification of coal seams and an example to a mechanical application in ELİ Darkale coal mine. *Proceedings of the 8th Coal Congress of Turkey,* Zonguldak, Turkey, pp. 31–51.

Chen, W., Shihao, T., Min, C., and Yong, Y. 2016. Optimal selection of a longwall mining method for a thin coal seam working face. *Arabian Journal for Science and Engineering,* 41(9):3771–3781.

Churth, E. S. 1981. Longwall mining of thin seams. *Conference on Ground Control in Mining,* West Virginia University, Morgantown, WV, pp. 239–259.

https://www.worldcoal.org/coal/uses-coal/coal-

http://www.viewpointmining.com/article/pinnacle-mine-breaks-world-record, taken on May 2018, Viewpoint Perspectives on Modern Mining Issue 11.

Myszkowski, M. and Paschedag, U. 2013. *Longwall Mining in Seams of Medium Thickness Comparison of Plow and Shearer Performance under Comparable Conditions.* Caterpillar, Peoria, IL, 51 p.

Myszkowski, M. and Bauckmann, S. 2014. *Operational Experiences with Automated Plow Systems at Pinnacle Mine in the USA.* Caterpillar Global Mining, Caterpillar, Peoria, IL, 11 p.

Paschedag, U. 2014. *Plow Technology, History and the State of the Industry.* ©2014 Caterpillar. Caterpillar Global Mining, Peoria, IL, 16 p.

Stopa, Z. and Myszkowski, M. 2015. Mining low coal seams, a Polish success. *World Coal,* March:1–6.

Wang, F., Tu, S., and Bai, Q. 2012. Practice and prospects of fully mechanized mining technology for thin coal seams in China. *Journal of the Southern African Institute of Mining and Metallurgy,* 112(2):161–170.

Westfalia Lünen, 1980. Longwall mining systems, catalogue.

13

Cutting Coal with Shearers, Some Application Examples

13.1 Introduction

Increasing demand on energy increases coal production requirement. According to 2017 data, around 37% of the world's electricity comes from coal (https://www.worldcoal.org/coal/uses-coal/coal-). The increased production in parallel to increase of the number of personnel leads to health and safety problems. Diminishing of coal reserves on surface forces the industry to exploit deeper underground coal reserves, which is also considered as a risk for health and safety. Around 64% of the world's anthracite and bituminous coal production came from underground mining in 2005 (Johnson 2007). Automated mechanized production is considered as the basic solution to supply the required high coal demand and reduce health and safety problems.

Shearers are the most common equipment for mechanized mining of coal by longwall method. Longwall mining requires driving one or more gateways (roadways) apart from each other and connected to each other by an in-seam gallery. The coal bed in between the two levels is produced along a panel, which is long enough. Longwall production might be either in retreat or advance system. Both coal excavation and roof support jobs are performed by mechanized ways in a fully mechanized longwall mining system, which highly improves productivity and safety. Excavation is basically performed by shearer-loaders, coal plows with a quite limited application, and roof support is performed by hydraulically powered roof supports (self-advancing). One of the jobs, either excavation or roof support, is performed as mechanized and the other is classically in semi-mechanized coal production systems. If both of the jobs are performed as classical ways, it is usually called

classical longwall production system. The levels of technology and automation require further developments to reach the final target of manless production systems.

The detailed explanation about the longwall mining method is given in
Chapter 11 of this book. Fully mechanized longwall applications in Imbat
Mining Inc. in Turkey producing lignite in a thick seam and Twentymile Coal
Mine in United States of America will be summarized as shearer applications
in this chapter.

13.2 Fully Mechanized Longwall Mining Application in Imbat Lignite Mine

Imbat Mining Inc. has been mining lignite in the city Manisa, Eynez prov-
ince of Soma district in the reserves of Turkish Coal Enterprises, by a contract
granted as royalty, since 2004. The company was applying classical longwall
before 2011, semi-mechanized longwall between 2011 and 2015, and fully
mechanized longwall since December 2015 (Ekici and Satilmis 2016). Total
lignite reserve of the Imbat lignite site is around 20 million tons. However,
the company was recently granted by another thick lignite mine with additional reserves nearby the existing one.

13.2.1 Geology of Soma Coal Basin and Imbat Lignite Mine

The Soma coal basin is one of the largest economic lignite-bearing alluvial
basins of western Turkey, with around 600 Mt of coal reserves dispersed
in 11 different areas (Gokmen et al. 1993). The Miocene succession of the
Soma coalfield contains three different lignite seams named, Lower Lignite,
Middle Lignite, and Upper Lignite. The Lower Lignite is a seam having an
average thickness of around 15 m–20 m with an average calorific value of
5,000 kcal/kg, but the Middle and Upper Lignite are successions comprised
of several lignite beds with thicknesses ranging from 0.15 m to 2.5 m with
calorific values less than 2,000 kcal/kg, inter-bedded with sandstone, mudstone, and siltstone (Inci 1998, 2002).

The lignite in Soma region is quite prone to spontaneous combustion.
Imbat Mining Inc. currently extracts lignite in Lower Lignite series, the
roof stone is marl. The dip of the lignite seam is around 5–10. A general stratigraphic section for the Soma coal basin is given in Figure 13.1
(Gokmen et al. 1993).

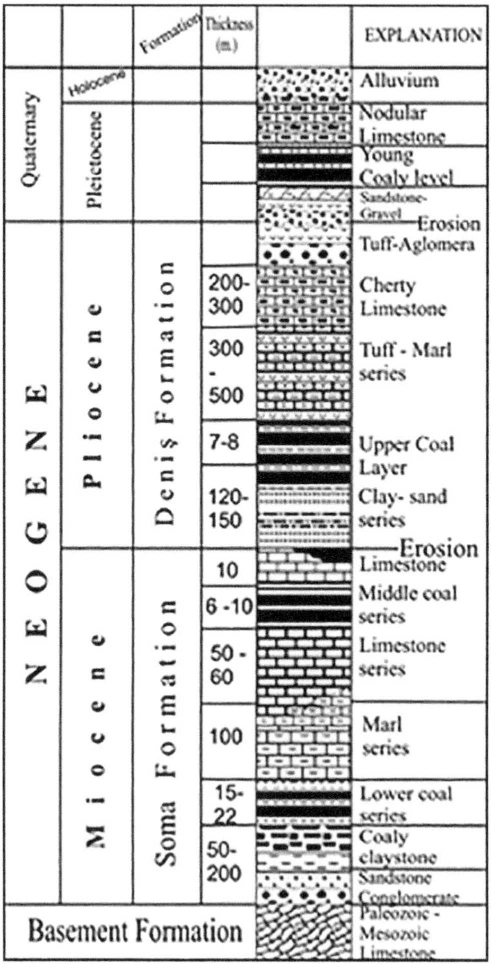

FIGURE 13.1
General stratigraphic section for Soma coal basin. (From Toprak, S., *Int. J. Coal. Geol.*, 7, 263–275, 2009.)

Imbat lignite mine has one basic fault aligning from north to south dividing the site into generally two halves, one on the east (~26 m of lignite seam) having a methane about 2.2 m³/ton and the other on the west (~12 m of lignite seam) as seen in Figure 13.2. The panels with different colors indicate different planned production years in Figure 13.2.

FIGURE 13.2
Plan view and longwall panels of Imbat lignite site. (With kind permission of Gokalp Buyukyildiz, Imbat Mining, Turkey.)

13.2.2 Physical, Chemical, and Mechanical Properties and Cuttability Characteristics of the Lignite and Bedrock in Imbat Lignite Mine

Detailed studies on physical, chemical, and mechanical properties and cuttability characteristics of the lignite and bedrock in Imbat lignite mine were performed by Bilgin et al. (2011) and published by Bilgin et al. (2015, 2016). All of these properties were also summarized in different chapters of this book (Chapters 5 through 7). Some of the chemical, physical-mechanical, and cuttability properties of the lignite seam are summarized

TABLE 13.1

Chemical and Heat Properties of the Coal Seam

Sample No.	Level (m)	Moisture (%)	Ash (%)	Volatile Matter (%)	Carbon (%)	Net calorific Value (kcal/kg)	Sulfur (%)
1	0.5	6.80	50.57	25.18	24.07	3,091	0.50
2	4.0	10.17	12.76	43.08	44.16	5,659	0.62
3	8.0	12.17	3.90	44.25	51.85	6,752	0.94
4	8.0	11.58	16.29	39.23	44.48	5,802	1.32
5	9.0	11.36	7.11	46.03	46.86	6,601	2.15
6	5.3	12.10	7.35	43.36	49.29	6,314	0.83
7	4.0	9.91	11.34	45.78	42.88	5,944	0.80
8	10.0	11.83	10.24	44.19	45.57	6,297	1.00
9	1.8	12.23	10.63	45.77	43.60	6,043	0.72
10	14.0	8.30	43.00	32.52	24.48	3,588	1.02
11	0.7	4.01	71.99	15.14	12.87	1,839	0.70
12	18.0	8.67	26.59	43.57	29.84	4,373	2.42
13	16.8	6.58	32.73	45.28	21.99	3,503	1.16
14	13.0	7.65	31.50	44.11	24.39	3,827	1.31
15	26.4	11.46	10.02	42.84	47.14	6,421	0.92
16	26.0	7.98	27.28	45.51	27.21	3,657	0.50
17	27.0	11.19	18.86	37.49	43.65	5,607	1.59
18	12.0	11.36	19.30	38.47	42.23	5,524	0.93
19	0.8	12.03	2.76	43.06	54.18	6,839	1.14
20	13.0	11.32	4.42	41.37	54.21	6,610	1.05
21	1.0	11.02	10.20	40.04	42.10	5,100	2.00
22	11.5	8.40	42.10	30.10	24.60	3,200	0.50
23	20.0	11.20	34.40	37.07	29.50	3,700	1.30
24	0.7	10.30	4.20	45.11	53.30	6,000	2.30
25	0.3	4.01	64.20	15.03	9.50	1,742	0.50
26	17.0	9.07	46.60	25.04	21.50	2,900	0.60
27	25.8	7.81	3.60	43.12	57.10	4,600	2.20
28	27.0	3.89	63.80	15.08	93.00	1,700	0.50

Source: Bilgin, N. et al., The cuttability and cavability of thick coal seam in Soma-Eynez coal field IR-75153 of TKI operated by IMBAT AS. Project report, Istanbul Technical University, Turkey, 59 p, 2011; Bilgin, N. et al., *Int. J. Rock Mech. Min. Sci.*, 73:123–129, 2015.

in Tables 13.1 through 13.3, respectively. The "Level" term in these tables indicates the distance from the roof of the seam. The roof stone marl has an impact strength index of 81, cone indenter hardness index of 2.54, uni-axial compressive strength of 64 MPa, Shore sclerescope index of 40, and 15.9 MJ/m^3 of specific energy (obtained by a chisel cutter tested in small scale linear cutting test set) (Bilgin et al. 2011).

TABLE 13.2

Mechanical Properties of the Coal Seam

Sample No.	Level (m)	Uniaxial Compressive Strength (MPa)	Impact Strength Index	Schmidt Hammer Rebound Hardness
1	0.5	89.1	74	28
2	4.0	38.5	79	59
3	8.0	19.2	45	30
4	8.0	26.1	76	64
5	9.9	5.9	72	34
6	5.3	14.1	72	25
7	4.0	8.2	74	45
8	10.0	7.1	74	46
9	1.8	19.4	73	34
10	14.0	8.4	69	34
11	0.7	55.7	73	56
12	18.0	14.4	71	34
13	16.8	32.0	72	40
14	13.0	7.8	70	37
15	26.4	14.0	71	35
16	26.0	35.0	71	51
17	27.0	36.1	71	75
18	12.0	26.0	62	27
19	0.8	49.6	76	30
20	13.0	17.9	75	43
21	1.0	28.3	75	30
22	11.5	24.0	74	30
23	20.0	18.8	61	33
24	0.7	19.4	73	45
25	0.3	55.7	73	48
26	17.0	40.1	67	50
27	25.8	14.0	71	31
28	27.0	24.0	72	54

Source: Bilgin, N. et al., The cuttability and cavability of thick coal seam in Soma-Eynez coal field IR-75153 of TKI operated by IMBAT AS. Project report, Istanbul Technical University, Turkey, 59 p, 2011; Bilgin, N. et al., *Int. J. Rock Mech. Min. Sci.*, 73:123–129, 2015.

TABLE 13.3

Cuttability Properties of the Coal Seam

Level (m)	FC (kN)	F'C (kN)	F'C/FC	FN (kN)	F'N (kN)	F'N/FN	FC/FN	SE (MJ/m³)
0.3	1.96	4.28	2.2	2.18	3.43	1.6	0.9	3.04
0.7	0.24	1.39	5.6	0.17	0.72	4.2	1.4	1.57
1.0	0.28	1.61	5.6	0.39	1.39	3.6	0.7	0.42
11.5	0.31	2.12	6.8	0.22	0.82	3.7	1.4	1.03
17.0	0.26	1.69	6.5	0.18	0.75	4.2	1.4	2.00
20.0	0.34	1.74	5.1	0.23	0.87	3.8	1.5	0.81
25.8	0.31	1.54	5.0	0.15	0.68	4.5	2.1	1.51
27.0	0.45	1.98	4.4	0.22	0.79	3.6	2.0	3.18

Source: Bilgin, N. et al., The cutt ability and cavability of thick coal seam in Soma-Eynez coal field IR-75153 of TKI operated by IMBAT AS. Project report, Istanbul Technical University, Turkey, 59 p, 2011; Bilgin, N. et al., *Int. J. Rock Mech. Min. Sci.,* 73:123–129, 2015.

FC: mean cutting force, F'C: mean peak cutting force, FN: mean normal force, F'N: mean peak normal force, and SE: mean specific energy.

13.2.3 Production Method and Equipment in Imbat Lignite Mine

Production in Imbat lignite mine is performed by fully mechanized retreat longwall mining with sublevel caving. Sublevel caving method is used with three consecutive and simultaneously working longwalls following each other with 25 m–30 m of distances (Figure 13.3). Having three production

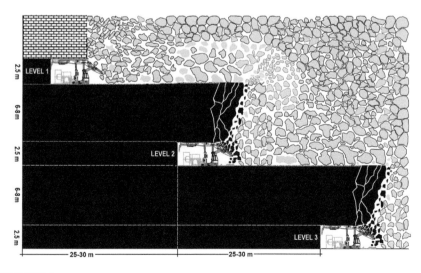

FIGURE 13.3
Longitudinal cross-section of a longwall panel of Imbat lignite mine. (With kind permission of Gokalp Buyukyildiz, Imbat Mining, Turkey.)

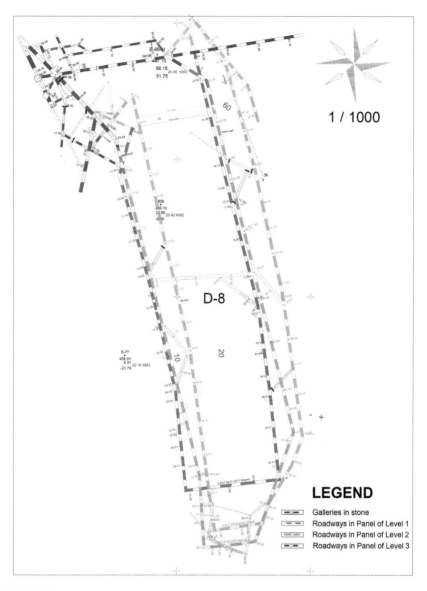

FIGURE 13.4
Roadway plan of the panel D8 of Imbat lignite mine. (With kind permission of Gokalp Buyukyildiz, Imbat Mining, Turkey.)

levels eases the methane evacuation from a panel. Panel lengths are usually around 800 m–1,000 m and face lengths are around 150 m. Mostly four panels are produced at the same time. Entrance to the mine is realized with four inclined galleries. Roadway plan of the panel D8 is presented in Figure 13.4. The mean overburden is around 600 m.

FIGURE 13.5
Photograph of the monorail used in Imbat lignite mine. (Shot by the authors, with kind permission of Gokalp Buyukyildiz, Imbat Mining, Turkey.)

The mean lignite production is around 20,000 tons/day, with an annual production of 6.0 Mt. Output lignite comes directly to coal preparation plant having a washing capacity of 600 tons/h.

The transport of lignite is realized by chain conveyors of 1,700 m within longwalls and conveyor belts of 7,700 m in roadways. The miners are transported via a belt conveyor of 1,230 m (with a width of 1.0 m) between the levels of +374 m and +40 m. Materials to and out of the mine are transported by six suspended monorails, of which four of them are diesel with toothed drive and two of them are diesel with frictional-toothed drive, and 14 frictional-toothed drive shunting trolleys powered by compressed air of 4.5 bar for loads up to 2 tons. A photograph of the monorail is given in Figure 13.5 and general technical properties of the monorails used in Imbat lignite mine are given in Table 13.4.

The heights of the faces in the levels 1, 2, and 3 are around 2.5 m–3.0 m. The level 1 (top slice or top face) contacts with the roof stone including 3 m–5 m of calcareous marl, and then marl and, therefore, there is only one armored face conveyor (chain conveyor, face conveyor) on the face of level 1. Drilling and blasting is used from time to time to ease caving the roof, especially when excavation starts newly in a panel. The sublevel between the levels

TABLE 13.4

Technical Properties of the Monorails Used in Imbat Lignite Mine (Becker Warkop Product Catalogue)

Brand; Type	Becker Warkop; BWTU-50/100
Rail with toothed rack cross-section	I 155 (I 140 E wg DIN)
Length of straight rails (standard)	1,980 mm
Toothed rack pitch	60 mm
Horizontal curvature profile of running line	Minimum 4 m
Vertical curvature profile of running line	Minimum 10 m
Angle of deviation from rectilinearity of suspension rails on joints in a vertical plane	Maximum ± 3°
Angle of deviation from rectilinearity of sleeve rails on joints in a vertical plane	0°
Angle of deviation from rectilinearity of suspension rails on joints in a horizontal plane	Maximum ± 0.5°
Angle of deviation from rectilinearity of sleeve rails on joints in a horizontal plane	0°
Permissible running line inclination	Maximum 45°
Load of rail joint in direction of suspension	200 kN
Load of rail joint along tracks	400 kN

1 and 2 (usually around 6 m–8 m height) caves itself and the caved lignite is drawn in the level 2. Since the floor stone is clay containing montmorillonite, which becomes highly plastic when it absorbs mine water, the bottom level of the face in the level 3 starts over the low calorific value lignite (~2.0 m thickness) located over the clay. The sublevel between the levels 2 and 3 (usually around 6 m–8 m height) caves itself and the caved lignite is drawn in the level 3. There are two chain conveyors in the levels 2 and 3, of which one (face conveyor with two chains on center) transports the lignite from the face, and the other (rear conveyor) transports the caved sublevel lignite behind the face. Around 95% of the coal is extracted/recovered in a panel, which is considered as very useful to reduce the spontaneous combustion risk of the lignite. Photographs of the face and rear conveyors are presented in Figures 13.6 and 13.7, respectively.

Plan view of a panel and equipment layout in level 2 are presented in Figure 13.8. The face conveyors each having a capacity of 500 tons/h are two-chained on center. The rear conveyors each having a capacity of 1,000 tons/h are 90° curved type having only one chain on center, which is the only application in the world other than China. All of the chain conveyors have a transportation speed of 1.0 m/s and width of 732 mm. The face conveyors have driving powers of 2 × 100 kW at maingate and tailgate. The rear conveyors have driving powers of 135 kW at tailgate and 400 kW at maingate. There are breakers (at the maingates) each having a

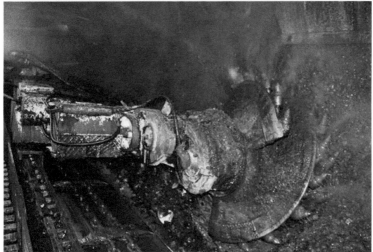

FIGURE 13.6
Photographs of the face conveyor used in Imbat lignite mine. (With kind permission of Gokalp Buyukyildiz, Imbat Mining, Turkey.)

capacity of 1,000 tons/h to reduce the size of big lignite chunks especially coming from the sublevel caving.

Lignite is excavated by double-ended ranging drum shearer-loaders with drum diameters of 1.4 m–1.6 m, drum powers of 2 × 160 kW, and webs of 0.70 m–0.85 m, providing for face heights from 2.6 m to 3.2 m. Conical cutting tools are used with water for dust spraying and tool cooling. The technical drawing of the shearer-loader is presented in Figure 13.9 for the positions at

FIGURE 13.7
Photographs of the rear conveyor used in Imbat lignite mine. (With kind permission of Gokalp
Buyukyildiz, Imbat Mining, Turkey.)

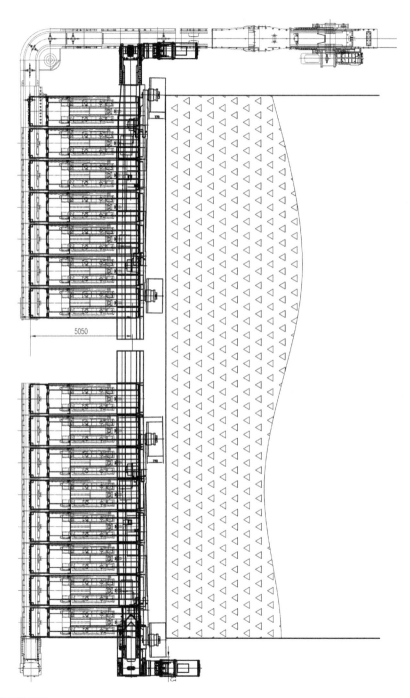

FIGURE 13.8
Plan view of a panel and equipment layout in Level 2, Imbat lignite mine. (With kind permission of Gokalp Buyukyildiz, Imbat Mining, Turkey.)

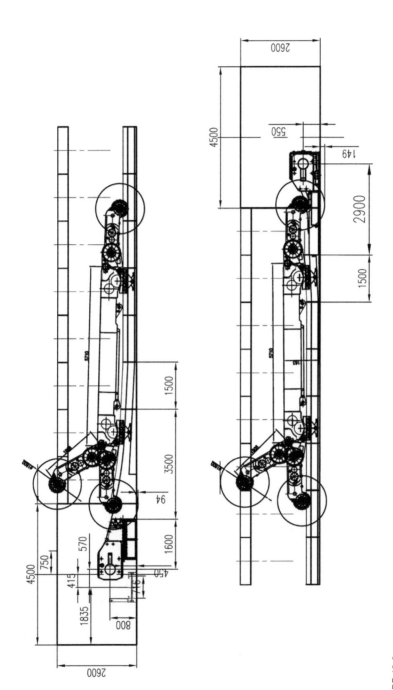

FIGURE 13.9
Technical drawing (cross-section) of the shearer-loader for positions at tailgate (top) and maingate (bottom). (With kind permission of Gokalp Buyukyildiz, Imbat, Turkey.)

FIGURE 13.10
Photographs of the shearer-loader. (With kind permission of Gokalp Buyukyildiz, Imbat Mining, Turkey.)

maingate and tailgate. Photographs of the shearer-loader are presented in Figure 13.10. The velocity of the shearer-loader is 6.3 m/min when it is not cutting, however, when cutting the lignite, a low velocity of around 1.5 m/min is applied to keep the machine breakdowns minimum and sublevel caving takes some time.

Roof support is provided by self-advancing powered supports each having six hydraulic props (Figure 13.11). Each set of powered supports having a mass of around 15.2 tons has a load capacity of 4,400 kN (~450 tons) and has a special hydraulically movable (back and forth) pistons at the rear

FIGURE 13.11
Photograph of the roof supports stocked in the site, Imbat lignite mine. (Shot by the authors, with kind permission of Gokalp Buyukyildiz, Imbat Mining, Turkey.)

end to break the large lignite chunks (coming from the sublevel caving) stacked between the rear end of the powered support and the rear conveyor (Figure 13.12). They have a minimum height of 1.75 m and a maximum height of 2.8 m, their widths are 1.0 m. They are able to push the face conveyor and pull the rear conveyor by hydraulic pistons.

The lignite in Imbat mine is very prone to spontaneous combustion. The fly ash obtained from a thermic power plant is mixed with water and pumped to the gob area to prevent spontaneous combustion of lignite. The average gas emission is around 2.2 m³/ton. Monitoring of the gas

FIGURE 13.12
Photograph of backside of the roof support to break the large lignite chunks. (Shot by the authors, with kind permission of Gokalp Buyukyildiz, Imbat Mining, Turkey.)

emissions has been realized with an automated/remotely controlled system since August 2009.

There are two separate ventilation circuits in the Imbat lignite mine, of which one is for the west panels and the other is for the east panels. The west panels are ventilated through two inclined galleries. The east panels are ventilated through two inclined galleries for air intake and one inclined gallery for air out. Suction type ventilation is applied in the site with totally four fans and seven spare fans, totally 11 fans having capacities from 1,200 m³/min to 3,000 m³/min and a total power of 1,875 kW.

Total water ingress in the Imbat lignite mine is around 80 m³/h, of which half of it comes from the ash-water mixture pumped into the gob area to prevent spontaneous combustion of lignite and equipment cooling water.

There are six roadheaders with different weight classes between 45 tons and 65 tons to excavate the galleries and roadways having cross-sections of 10 m²–16 m² in horseshoe or rectangular shapes. The roadways within the coal requires trimming and re-supporting due to ground squeezing or heavy deformation.

Productivity in the fully mechanized longwall mining in Imbat lignite site is around 53.3 tons/person/day (Ekici and Satilmis 2016).

13.3 Fully Mechanized Longwall Mining Application in Twentymile Thermal Coal Mine

This section is summarized after Nettleton and Berdine (2013). Twentymile Mine is located in Northwest Colorado by Peabody Energy. Historically, operations transitioned from surface to underground mining, with the initiation of continuous miner operations in 1983, and from room and pillar to longwall mining in 1989. The coal production was also gradually expanded up to 8.5 Mt/year. The current longwall system in the mine has a production capacity of 5,000 t/h with a face length of around 305 m and panel length of around 3,000 m–5,000 m. There are currently 430 employees working in the mine. Twentymile Mine is the state's largest coal producer and is consistently one of the safest mines in Colorado and the United States of America.

Twentymile's operations are located in the Yampa coalfield of the Uinta coal basin, and primary production is from the Wadge coal seam. The Wadge seam averages 11,200 Btu, 9% ash, 9% moisture, less than 0.5% sulfur, and very low levels of mercury. The Wadge seam occurs within a shallow basin, is relatively flat-lying (dipping to the north at 9°), and has a consistent thickness over the property of approximately 2.9 m. Overburden is a sequence of shale, claystone, and sandstone units deposited in marine, shoreline, and near-shore environments, and ranges from 300 m to 530 m thick.

Continuous miners are used for development operations. Twentymile's continuous mining equipment includes Joy 12CM12 and 12CM27 miners, Fletcher dual-boom roof bolters, Joy 10SC32 shuttle cars, and Stamler feeder breakers. The longwall development sections are typically a three-entry design. Resin-anchored roof bolts (1.8 m–2.4 m), cable trusses, and roof mats are used throughout the mine to provide supplemental support.

The current longwall system consists of a 150 Caterpillar shields of 2.05-m height, a Joy 7LS5 shearing machine, and Caterpillar-supplied automated face conveyor, stage loader, crusher, and tailpiece. The longwall face is 305 m wide, panels are up to 5,000 m long, and each cut removes 1 m (web) of coal from the 2.9-m coal seam. On a typical operating shift, the shearer achieves a typical production rate of 2,500 tons/h–3,500 tons/h.

The shearer is equipped with 740-kW (1,000-hp) ranging arms. The face conveyor runs at a speed of 110 m/min and is powered with three 880-kW (1,200-hp) motors. The stage loader is powered by a 740-kW (1,000-hp) unit operating at a speed of 140 m/min. Power is supplied to the longwall system at 4,160 V.

A scissorveyor (collapsible belt unit), designed by Twentymile and supplied by Continental Conveyor, allows the longwall to advance 58 m before removing belt structure. Five workers can complete removal of 58 m of belt structure to advance the scissorveyor on a maintenance shift in 5 h.

A series of 1.8-m-wide conveyors, supplied by Continental Conveyor, are used to haul coal to the surface at a speed of 4 m/s. Conveyor design capacity is 5,000 tons/h (continuous), and an underground surge bunker provides some buffering capacity.

The cutting sequence is designed to cut the center out of the seam to relieve the horizontal stresses, and then to cut the top and bottom of the seam. The center of the seam is cut with the shearer traveling from the headgate to the tailgate. The thickness of the initial web cut varies with conditions to maximize production rates. The typical web cut removes 50%–70% of the web depth. As the shearer approaches the tailgate, it cuts the full seam thickness beginning 19 shields from the end. The shearer returns to the headgate cutting the remaining 30%–50% of the web and the top and bottom of the cut. The shearer has a maximum speed of 43 m/min.

This method allows for high production rates and has the added advantage of sequencing shield advance so that any dust generated by shield movement is in the return airflow, "downwind" of the shearer operators.

13.4 Numerical Example: Estimation of Adequacy of a Shearer

Check the power and haulage force requirements of the shearer-loader selected for excavation of a thick coal seam. The technical specifications of the shearer-loader are given in Table 13.5.

It is assumed that the speed of drum rotation (RPM) is 25 rpm and the required haulage speed (H_S) is 1.5 m/min, as in the case of application in Imbat lignite mine. Full-scale linear cutting experiments performed on blocks of coal samples by using a conical cutter at single scroll/start cutting pattern and different line spacings and depths of cut indicate that the optimum line spacing to depth of cut ratio $(s_L/d)_{opt}$ is around 1.5. Although a triple-spiral cutting pattern is applied in the shearer-loader, it is known that there is not much difference in terms of the forces acting on a tool between single- and triple-spiral cutting patterns (Copur et al. 2017). Cutting (FC) and normal (FN) forces acting on the tools by cutting

TABLE 13.5

Technical Properties of the Selected Shearer-Loader

Technical Properties	
Type of shearer loader	Double Drum Ranging Arm
Cutting height	3.2 m
Drum cutting diameter	1.6 m
Web width including clearance plate	0.84 m
Maximum speed of drum rotation	25 rpm
Haulage speed capacity	0 m/min–6.5 m/min
Haulage force capacity	600 kN
Cutting power capacity	2 × 150 kW
Haulage power capacity	2 × 50 kW
Total mass	800 kN
Assumptions on Drum Design	
Tool type	Conical tools
Tool reach for 48° attack angle	6 cm
Vane (start, spiral, sequence) number	3 vanes
Tool number per cutting line	1 tool/line
Line spacing of tools	3 cm
Cut spacing of tools	9 cm

perpendicular to the cleat planes are given in Equations (13.1) and (13.2) depending on the depth of cut (per tool per revolution) for a line spacing of 3 cm:

$$FC = 2.2 \cdot d^{0.5} \qquad (13.1)$$

$$FN = 1.1 \cdot d^{0.7} \qquad (13.2)$$

FC and FN are in (kN) and d is in (cm) in these equations.

Solution:

Advance per revolution of the drum is estimated to be 6 cm/rev by using Equation (13.3):

$$d_{rev} = H_S / RPM \qquad (13.3)$$

where,

d_{rev} = advance per revolution of the drum (cm/rev),
H_S = haulage speed of the shearer-loader (cm/min), and
RPM = rotational speed of the drum (revolution/min, rpm).

Since there are three vanes (starts, spirals, sequences) in one tool per cutting line system on a drum, maximum penetration of each sequence at mid-drum level for each revolution of the drum ($d_{max} = d_{vane} = d_{spiral}$) is estimated to be 2 cm/tool/rev by using Equation (13.4):

$$d_{max} = d_{vane} = d_{spiral} = d_{rev} / V_N \tag{13.4}$$

where, V_N = number of vanes on a drum.

Shape of the area cut by a tool, as the drum rotates and moves forward, approximates to a cissoid (crescent). The depth of cut (penetration) of a tool is almost zero at the beginning of a cut. It becomes maximum, when the tool reaches at mid-drum level. The maximum depth of cut at mid-drum height for this problem is 2 cm/tool/rev. The average depth of cut (d_{ave}) can be calculated as 0.64 of the maximum depth of cut (Roxborough and Phillips 1981), which gives 1.3 cm/rev of average depth of cut. For this depth of cut value, FC is estimated to be 2.5 kN and FN is estimated to be 1.3 kN by using Equations (13.1) and (13.2).

Since the drum width (web, D_W) is 84 cm, the number of vane tools (N_{VT}) on a drum is estimated to be 28 tools by using Equation (13.5):

$$N_{VT} = D_W / s_L \tag{13.5}$$

If it is assumed that the number of tools on the face plate/clearance ring is nine, then, the total number of tools (N_{TT}) on a drum is estimated to be around 37 tools by using Equation (13.6):

$$N_{TT} = N_{VT} + N_{FPT} \tag{13.6}$$

where, N_{FPT} = number of the tools mounted on face plate.

Cutting power of a drum ($P_{CUT\text{-}NET}$) is estimated by using Equation (13.7):

$$P_{CUT\text{-}NET} = 2 \cdot \pi \cdot N \cdot T \tag{13.7}$$

where,

N = drum rotational speed per second (RPM/60) and
T = torque requirement of drums for cutting (kNm).

Torque (T) is estimated by using Equation (13.8):

$$T = N_{TC} \cdot FC \cdot R_D \tag{13.8}$$

where,

N_{TC} = number of tools in contact with coal,
FC = cutting force acting on a tool (kN), and
R_D = radius of the drum (m).

Since the half of a drum sumps into the coal, the number of tools in contact with coal is $0.5 \times N_{TT} = 19$ tools per drum for the selected problem. Since there are two ranging drums on the shearer-loader, (N_{TC}) is estimated to be $2 \times 19 = 38$ tools. Drum radius is $R_D = D_D/2 = 1.6/2 = 0.8$ m. It is assumed that the tools placed on face plate/clearance ring are closely spaced so that the tool forces acting on them are almost equal to the forces acting on the vane tools.

Based on these considerations and estimations, the net cutting power of two drums ($P_{CUT\text{-}NET}$) is estimated as 200 kW by using Equation (13.9):

$$P_{CUT\text{-}NET} = 2 \cdot \pi \cdot N \cdot T = 2 \cdot \pi \cdot (RPM/60) \cdot N_{TC} \cdot FC \cdot R_D$$

$$P_{CUT\text{-}NET} = 2 \cdot \pi \cdot \left(\frac{25}{60}\right) \cdot 38 \cdot 2.5 \cdot 0.8 \cong 200 \frac{kNm}{s} = 200 \text{ kW} \tag{13.9}$$

If the efficiency of the drum power transmission is assumed to be 0.8, the output (gross) power of the two drums for cutting should be ($200/0.8 = 250$ kW $= 2 \times 125$ kW), which is lesser than the installed cutting power of the drums (300 kW $= 2 \times 150$ kW). This means that the powers of the drums are good enough with a certain safety margin to achieve the required performance.

Net power requirement of the shearer-loader for pulling over the armored face conveyor ($P_{PULL\text{-}NET}$) is estimated by using Equation (13.10):

$$P_{PULL\text{-}NET} = P_{THRUST\text{-}NET} + P_{MOVE\text{-}NET} \tag{13.10}$$

where,

$$P_{THRUST\text{-}NET} = N_{TC} \cdot FN \cdot (H_S/60) \tag{13.11}$$

$$P_{MOVE\text{-}NET} = W \cdot f \cdot (H_S/60) \tag{13.12}$$

where,

$P_{THRUST\text{-}NET}$ = force required for penetration of the cutters into coal (kN),
$P_{MOVE\text{-}NET}$ = force required to move the shearer-loader (kN),
W = force acting on the mass center of the shearer-loader (kN), and
f = friction coefficient between the shearer-loader and the armored face conveyor (which can be assumed to be 0.3 for this problem).

Therefore,
$P_{THRUST\text{-}NET} = 38 \times 1.3 \times (1.5/60) = 1.2$ kW,
$P_{MOVE\text{-}NET} = 800 \times 0.3 \times (1.5/60) = 6.0$ kW, and
$P_{PULL\text{-}NET} = 1.2 + 6.0 = 7.2$ kW.

If the efficiency of the power transmission of haulage motors is assumed to be 0.8, the output power for haulage should be (7.2/0.8 =) 9.0 kW, which is well below the installed haulage power of 50 kW, which means that the haulage power of the shearer-loader is good enough to achieve the required performance. Power requirement of the vanes for loading is ignored in this study, since it is quite small (1%–2% of the total) compared to power requirement for cutting.

Total net power requirement of the shearer-loader ($P_{TOTAL-NET}$) is estimated to be 209 kW by using Equation (13.13):

$$P_{TOTAL-NET} = P_{CUT-NET} + P_{PULL-NET} \tag{13.13}$$

The haulage force acting on the shearer loader (FH) can be estimated to be 290 kN by using Equation (13.14), which is lower than the haulage force capacity of the shearer-loader (600 kN). This means that the haulage force capacity of the shearer-loader is good enough to achieve the required performance.

$$FH = N_{TC} \cdot FN + W \cdot f \tag{13.14}$$

This special example does not include the effect of any particular operational conditions such as working uphill or downhill, geological irregularities on the floor, partings, etc. and only accounts for rough estimation of power requirement of a shearer-loader. Loading capacity of the vanes should also be checked. Detailed analysis of force balance of the drums should also be analyzed by computer simulations of drums.

13.5 Conclusive Remarks

The representatives of Imbat Mining Inc. applied all types of longwall mining system in their operations including in the order of classical, semi-mechanized, and, finally, fully mechanized.

The Twentymile Mine is one of the most technologically advanced mines in the world, but it is how its people use the technology that makes the difference in safety, productivity, and costs.

A systematic approach to planning and management maximizes organizational effectiveness and fosters continuous improvement. The key factor in its success is teamwork, with employee involvement at every level of the process.

It should also be added that the rules of cutting mechanics are a useful and necessary tool for properly designing and selecting shearers for efficient production.

References

Becker Warkop. Product Catalogue. http://www.becker-mining.com.pl/files/pdf/transporten.pdf.

Bilgin, N., Balci, C., Copur, H., Tumac, D., and Avunduk, E. 2015. Cuttability of coal from the Soma coalfield in Turkey. *International Journal of Rock Mechanics and Mining Sciences,* 73:123–129.

Bilgin, N., Copur, H., and Balci, C. 2016. Use of Schmidt hammer with special reference to strength reduction factor related to cleat presence in a coal mine in Soma,Turkey. *International Journal of Rock Mechanics and Mining Sciences,* 84:25–33.

Bilgin, N., Copur, H., Balci, C., Avunduk, E., and Tumac, D. 2011. The cuttability and cavability of thick coal seam in Soma-Eynez coal field IR-75153 of TKI operated by IMBAT AS. Project report, Istanbul Technical University, Turkey, 59 p.

Copur, H., Bilgin, N., Balci, C., Tumac, D., and Avunduk, E. 2017. Effects of different cutting patterns and experimental conditions on the performance of a conical drag tool. *Rock Mechanics and Rock Engineering,* 50(6):1585–1609.

Ekici, A. and Satilmis, U. 2016. Investigation and comparison of the semi and fully mechanized systems applied at the Imbat Mining Company. *Proceedings of the 20th Coal Congress of Turkey,* May 4–6, Zonguldak, Turkey, pp. 11–24 (in Turkish).

Gokmen, V., Memikoglu, O., Dagli, M., Oz, D., and Tuncali, E. 1993. Inventory of lignites in Turkey. General Directorate of Mineral Research and Exploration (MTA), Ankara, Ture, Turkey, 355 pp (in Turkish).

https://www.worldcoal.org/coal/uses-coal/coal-.

Inci, U. 1998. Lignite and carbonate deposition in middle lignite succession of the Soma formation, Soma coalfield, western Turkey. *International Journal of Coal Geology,* 37(3–4):287–313.

Inci, U. 2002. Depositional evolution of Miocene coal successions in the Soma coal-field, western Turkey. *International Journal of Coal Geology,* 51(1):1–29.

Johnson, D. 2007. *Latest Technology and Performances in the Global Underground Coal Mining Industry.* Joy Mining Machinery, Milwaukee, WI, 38 p.

Nettleton, J. and Berdine, M. 2013. *Modern American Coal Mining: Methods and Applications.* Section 12: Case Studies. Case Study 9: Twentymile Mine. C.J. Bise Editor, Published by the Society for Mining, Metallurgy, and Exploration, Englewood, CO. Ebook 978-0-87335-395-3.

Roxborough, F. F. and Phillips, H. R. 1981. *Applied Rock and Coal Cutting Mechanics.* Workshop Course 156/81, Australian Mineral Foundation, Adelaide, Australian. 11–15 May.

Toprak, S. 2009. Petrographic properties of major coal seams in Turkey and their formation. *International Journal of Coal Geology,* 78(4):263–275.

14

Cuttability Criteria for the Use of Continuous Surface Miners and Worldwide Applications in Surface Coal Mining

14.1 Introduction

The need of using a selective mining technology emerged at the beginning of 1983 to increase mine productivity and efficiency compared to conventional mining by excavating each layer in the coal seam separately.

Volk (2016) stated that there were around 484 surface continuous miners (Wirtgen) in the world per April 15, 2015 working in mineral and coal productions. He emphasized the advantages of this technology as: higher unit production, selective mining of each layer with reduced loss and dilution, increased amount of target size material and reduced amount of fines, excavating siderite intrusions in coal with normal mining operations, the resultant smaller grain size having positive impact to the wash plant, friendly mining causing less environment problems, increased yield of coal handling, and preparation plant.

The main objective of this chapter is to discuss the main features of this environmentally friendly mining method in relation to the selection criteria of continuous surface miners based on the geotechnical properties of coal and coal measure rocks. Some examples of worldwide applications are also given.

14.2 Working Principles of Continuous Surface Miners and Some Technical Characteristics

A schematic view of a surface continuous miner is given in Figure 14.1. It is a three or four track-mounted machine. It essentially comprises a cutting drum, disposing unit, and a working unit. It is powered by a diesel engine,

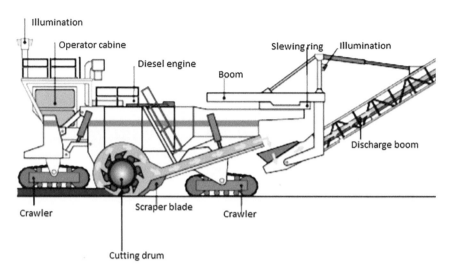

FIGURE 14.1
A schematic view of a continuous surface miner. (From Amar, P., *Int. J. Min. Sci. Technol.*, 23, 33–40, 2013.)

and hydraulic pumps are used for transmitting the power for different operations. The cutting drum, which is ideally located at the center of the gravity of the machine, is usually fitted with conical cutters and cuts and loads the material onto the primary conveyor. The drum rotates in and up in cutting direction, and a layer of predetermined thickness of the seam is cut and crushed to the optimal size by the cutting tools. Depth of cut is controlled by the "electronic depth control system" which enables the cutting drum to achieve the preset cutting depth. Depth of cut is important for determining the theoretical output (depth of cut × drum width × velocity of movement) of a surface miner (Bilgin et al. 2014, Dey and Ghose 2011).

The use of a surface miner leads to a vital change of the mining operations. A surface miner replaces the conventional mining steps of drilling and blasting or ripping and subsequently loading. Additionally, the primary crushing circuit is eliminated. The particle size distribution enables a better truck utilization as well as introducing the chance of transporting the material directly into conveyors. Especially in thin seams having a thickness of less than 2 m, application of a surface miner provides an improved selectivity compared to conventional mining methods (Bilgin and Balci 1996, Volk 2016).

Operational parameters of a continuous miner enable of controlling the size of the end product. High levels of coal fines in the run of mine material result in higher costs for washing, lower recovery, reduced workplace safety, and negative impacts on the downstream processes. Knowing where these coal fines come from and optimizing the procedures along the process chain can yield significant savings. The effect of the particle size on the quality of the end product is given in Table 14.1.

TABLE 14.1

Processing of Typical Particle Size Ranges in Coal

Particle Size Range	Processed In	Rating
>40 mm	Primary/secondary crusher	Ok, but additional costs and fines generation
2 mm–40 mm	Coarse circuit	Target size material, easy to process
0.075 mm–2 mm	Fine circuit	Ok, but more expensive to process
<0.075 mm	Loss	Undesirable

Source: https://www.wirtgen-group.com/en/news-media/press-releases/wirtgen-surface-mining-maximizing-coalrecovery-by-minimizing-fines.106243.php

There are different manufacturers of continuous surface miners in the market. Technical characteristics of some machines are given in Table 14.2 (Dutta 2012). Some of these machines have cutting drums in the middle of the machines, and some others have cutting drums mounted on a boom in front of the machine.

TABLE 14.2

Technical Characteristics of Continuous Surface Miners

	Parameters	Drum Width (m)	Machine Power (kW)	Operating Weight (ton)	Rated Weight (m³/h)	Cutting Depth (mm)	Maximum Cutting Speed (m/min)
Wirtgen GmbH	SM2100	2.0	448	41	550	250	25
	SM2200	2.2	671	49	668	350	84
	SM2500	2.5	783	100	845	600	25
	SM3500	3.5	895	137	1900	470	25
	SM4200	4.2	1194	184	2400	600	20
Vermeer	T855	2.5	281	40.8	NA	812	28
	T955	3.4	309	56.7	NA	812	20
	T1055	3.4	317	61.2	NA	812	16
	T1255	3.7	447	99.8	NA	610	12
L & T	KSM223	2.2	597	NA	NA	350	83
	KSM304	3.0	895	100	NA	400	20
TAKRAF GmbH	MTS180	3.3	500	NA	180	700	NA
	MTS300	4.0	750	NA	300	875	NA
	MTS500	4.9	1650	NA	500	1050	NA
	MTS800	5.6	2000	NA	800	1225	NA
	MTS1250	6.5	2500	NA	1250	1400	NA
	MTS2000	7.4	2500	NA	2000	1575	NA
Bitelli	SF202	2.0	515	43	180	250	NA

Source: Dutta, S., Application of surface miner in Indian coal mines, BSc. thesis, Department of Mining Engineering, National Institute of Technology Rourkela, Odisha, 56 p, 2012.

14.3 Some Examples of the Worldwide Applications of Continuous Surface Miners in Coal Mining

Some production rates of continuous surface miners obtained in (ton/h) in different countries are tabulated in Table 14.3. However, it should be emphasized that the production rates depend on the power of the machine, geological and geotechnical characteristics of the strata, skill of the operator, maintenance program, the job organization, and so on. Some of these factors will be discussed in the following section.

14.4 Cuttability Characteristics of Coal Seams and Selection Criteria Based on Geotechnical Properties of Coal and Power of Continuous Surface Miners

The cuttability characteristics of coal seams in respect to selection of continuous surface miners and prediction of the production rates are not easy tasks, which are investigated in detail by several authors.

Dey and Ghose (2008, 2011) reviewed the existing cuttability indices and proposed a new cuttability index, which is the sum of the rating of five parameters: point load index, volumetric joint count, Cerchar abrasivity index, direction of cutting respect to major joint direction, and machine power. Rating of the parameters of this rock mass cuttability classification is tabulated in Table 14.4. Assessment of excavatability by surface miners based on cuttability index is given in Table 14.5.

TABLE 14.3

Some Production Rates Obtained by Continuous Surface Miners in Different Countries

Machine	Country	Mine	Year	Production (t/h)	References
Wirtgen 3000SM	Australia	Western Collieries	1988	Max. 600 Mean 247	Yazici et al. (1998)
Wirtgen 3500SM	Bosna	Gacko	1990	803–1185	Yazici et al. (1998)
Wirtgen 3500SM	Australia	Mount Thorley	1990	1330–1600	Yazici et al. (1998)
Wirtgen 2200	India	Lakhanpour	2012	198	Dutta (2012)
Wirtgen 2200	India	Basundhara	2102	139	Dutta (2012)
Wirtgen 2100	India	Lakhanpur	2012	210	Dutta (2012)
Wirtgen 2200	China	Shengli	2016	895	Wirtgen (2016)
Wirtgen 4200	USA	Red Hills	2016	2800	Wirtgen (2016)
Wirtgen 4200	Australia	Australia New Acland	2016	3000	Wirtgen (2016)

TABLE 14.4

Rating of the Parameters of Rock Mass Cuttability Classification for Continuous Surface Miners

Class	I	II	III	IV	V
Point load index (I_s50)	<0.5	0.5–1.5	1.5–2.0	2.0–3.5	>3.5
Rating (I_s)	5	10	15	20	25
Volumetric joint count	>30	30–10	10–3	3–1	1
(number/m³)	5	10	15	20	25
Rating (J_v)					
Cerchar Abrasivity index	<0.5	0.5–1.0	1.0–2.0	2.0–3.0	>3.0
Rating (A_w)	3	6	9	12	15
Direction of cutting respect	72–90	54–72	36–54	18–36	0–18
to major joint direction	3	6	9	12	15
Rating (J_s)					
Machine power (kW)	<1000	800–1000	600–800	400–600	>400
Rating (M)	4	8	12	16	20

Source: Dey, K. and Ghose, A.K., *J. Mines Met. Fuels.*, 56, 85–91, 2008; Dey, K. and Ghose, A.K., *Rock Mech. Rock Eng.*, 44, 601–611, 2011.

TABLE 14.5

Assessment of Excavability for Continuous Surface Miners Based on Cuttability Index

Excavability Index (CI)	Possibility of Ripping
50 > CI	Very easy excavation
50 < CI < 60	Easy excavation
60 < CI < 70	Economic excavation
70 < CI < 80	Difficult excavation, may be not economic
CI > 80	Surface miner should not be deployed

Source: Dey, K. and Ghose, A.K., *J. Mines Met. Fuels.*, 56, 85–91, 2008; Dey, K. and Ghose, A.K., *Rock Mech. Rock Eng.*, 44, 601–611, 2011.

A relationship was also developed to predict the production rates of surface miners using the proposed index (Dey and Ghose 2008, 2011). Production rate of a surface miner can be estimated by the same authors as:

$$L = \left(1 - \frac{CI}{100}\right) \cdot k \cdot M_c \tag{14.1}$$

where,

L = production or cutting performance (bcm/h);

M_c = rated capacity of the machine (bcm/h);

CI = cuttability index; and

k = a factor for taking into consideration the influence of specific cutting conditions and is a function of pick lacing (array), pick shape, atmospheric conditions, so on, and varies from 0.5 to 1.0.

TABLE 14.6

Performance Analysis of Continuous Surface Miners for Three Seams

Parameter	Coal	Coal Gray Shale Patch	Shally Coal
Point load index	1.1	2.5	2.2
I_s	10	17	15
Volumetric joint count	32	20	33
J_v	5	8	5
Cerchar abrasivity	0.4	1	0.4
A_w	3	6	3
Machine power (kW)	448	448	671
M_c	16	16	10
Direction of machine operation with respect to joint plane (degree)	80	80	80
J_s	3	3	3
Cuttability index (CI)	37	51	36
Possibility of cutting	Very easy	Easy	Very easy
Density (t/m³)	1.6	1.9	1.6
Cutting condition	Poor	Poor	Poor
K	0.6	0.6	0.6
Rated machine capacity (m³/h)	400	400	668
Expected production achieved (tonne/h)	243	223	410
Actual production achieved (tonne /h)	225	160	394

Source: Dey, K. and Ghose, A.K., *J. Mines Met. Fuels.*, 56, 85–91, 2008; Dey, K. and Ghose, A.K., *Rock Mech. Rock Eng.*, 44, 601–611, 2011.

Based on the new cuttability index, Dey and Ghose (2018, 2011) compared the field production with the predicted production as given in Table 14.6. As seen from this table, the predicted production and the field production are very close to each other.

Origliasso et al. (2014) used the manufacturers' and experimental data, and they proposed a new method to predict both the production rate (PR) and the cutting speed of continuous surface miners. They found that uniaxial compressive strength (UCS) and Cerchar abrasivity index (CAI) of rock and the machine's engine power (P_w) were the three most important factors that influence the production rate (PR) as given in Equation 14.2.

$$PR = \left(2 \cdot P_w - 600\right) \cdot e^{-0.024 \cdot \left(UCS + 10 \cdot \left(CAI - 0.5\right)\right)} \tag{14.2}$$

Moreover, the cutting depth, UCS, and P_w had a significant impact on the cutting speed. They also proposed a new method and equations to determine the energy required to cut a unit volume of rock, for example, specific energy (SE) and established the relationship between SE, UCS, and PR. It was suggested that the results of this study could be used by practicing engineers to evaluate the applicability of the machines to a specific mine site.

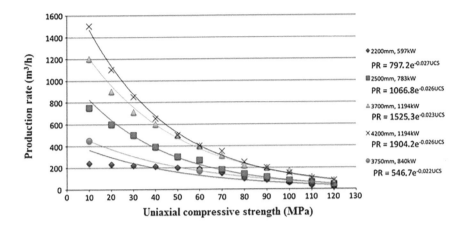

FIGURE 14.2
Variation of production rate with uniaxial compressive strength of the strata. (From Origliasso, C. et al., *Rock Mech. Rock Eng.*, 47, 757–770, 2014.)

The same authors revaluated also the information given by Wirtgen for handy use as given in Figure 14.2 that shows the variation of production rate with UCS of the strata.

Prakash et al. (2013) analyzed different field data from Indian mines and limestone quarries, equipment models, operating methods, and cutting performance assessment models. They found that production was related linearly to engine hour meter reading, diesel, and pick consumptions. Typical daily productions for coal mines were up to 4995 m³/day with average pick consumption of around 8. The average diesel consumption for each cubic meter of coal production was found to be around 0.26 L.

14.5 An Example from Turkey for Application of Continuous Surface Miners

14.5.1 General Information on the Studied Area and Mechanical/ Cuttability Characteristics of the Coal Seams

A laboratory coal cutting program was carried out in Istanbul Technical University (ITU) Mining Engineering Rock Excavation Laboratories to see the feasibility of using a continuous surface miner in Milten coal fields in Karaburun area in Istanbul (Yazici et al. 1998). Oligocene-aged coal seams consisted of Upper seam, Yuzluk seam, and Altmislik seam. Thin clay bands existing within the coal seams decreased calorific values of the end product. The coal mined in this area was intended for domestic use and increasing

TABLE 14.7

Some of the Mechanical Properties of the Three Coal Seams Subjected to Full-Scale Linear Cutting Tests

Coal Seam	Compressive Strength (MPa +/– st.dev.)	Impact Strength Index (+/– st.dev.)
Upper seam	33.6 +/– 1.4	85 +/– 1.0
Yuzluk seam	37.7 +/– 1.5	82 +/– 1.7
Altmislik seam	44.4 +/– 5.2	79 +/– 1.2

Source: Bilgin, N. and Balci, C., New excavation technologies in aggregate industry, *Proceedings of the 1th National Symposium on Aggregate Industry,* Turkish Chamber of Mines, Istanbul Branch, Turkey, 207–225, 1996.

the calorific value of the end product was of prime importance by selective mining operations. The mine is currently abandoned and one of the biggest airports in the world is being constructed in this area. Some of the mechanical properties of the three coal seams are given in Table 14.7.

The relationships between mean cutting force, FC (kN), mean normal force, NF (kN), and depth of cut, d (mm), for the hardest coal seam are given in Equations 14.3 and 14.4.

$$FC = 0.36 \cdot d \qquad (14.3)$$

$$FN = 0.37 \cdot d \qquad (14.4)$$

However, it should be mentioned here that cutting depth in these equations is not cutting depth of the drum, but the cutting depth of the cutter.

The relationship between SE and cutter spacing/cutting depth ratio (s/d) is one of the most important parameters in designing cutting drums. This relation is given in Figure 14.3 for Milten coal mine.

14.5.2 Selecting the Most Appropriate Surface Continuous Miner for Given Conditions

As seen from Figure 14.3, optimum s/d ratio was obtained at around 3 (Yazici et al. 1988). SE decreased with depth of cut and leveled off when depth of cut reached to 1 cm, in that case, optimum cutter spacing would be 3 cm. If the width of cutting drum is taken as 2.2 m, the number of cutters should be 220/3 = 73 cutters. If six gauge cutters are taken, the number of conical cutters must be 79.

The power of the cutting drum may be calculated using Equation 14.5:

$$P = 2 \cdot \pi \cdot N \cdot T \qquad (14.5)$$

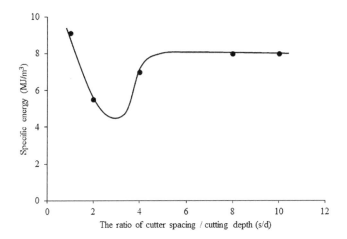

FIGURE 14.3
Relationship between specific energy and cutter spacing/cutting depth ratio for Milten coal mine, Turkey.

In the above equation, the parameters are as given below:
  P = Power of the cutting drum in kW;
  N = Number of revolution of the cutting drum per second, and it is usually 20 rpm; and
  T = Torque of the cutting drum in kNm.

Torque is the product of the number of cutters, the radius of cutting drum, and cutting force acting on a cutter. Cutting force for one cutter is calculated by using equation 14.3 as 3.6 kN for 10 mm depth of cut. Thus, torque of the cutting drum comes to be as:

$$2 \cdot 3.14 \cdot (20/60) \cdot 79 \cdot (2.2/2) \cdot 3.6 = 656 \text{ kW or } 880 \text{ hp}$$

The estimated power for the given condition is higher than that of SM 2200, therefore, one bigger version of this machine is proposed.

14.6 Conclusive Remarks

The use of continuous surface miners increases worldwide day by day. The main advantages of these machines are higher unit production, selective mining of each layer with reduced loss and dilution, increased amount of target size material, and reduced amount of fines. Volk (2016) stated that there were around 484 surface continuous miners (Wirtgen) in the world per April 15, 2015.

The use of a surface miner leads to a vital change of the mining operations. A surface miner replaces the conventional mining steps of drilling and blasting or ripping and subsequently loading. Additionally, the primary crushing circuit is eliminated. The particle size distribution enables a better truck utilization, as well as introducing the chance of transporting the material directly into conveyors. Especially in thin seams having a thickness of less than 2 m, the application of a surface miner provides an improved selectivity compared to conventional mining methods.

It should be emphasized that the production rates depend on the power of the machine, geological and geotechnical characteristics of the strata, skill of the operator, maintenance program, the job organization, and so on. One of the most important factors within these parameters is certainly the cuttability of the strata that is not an easy task to define, which is investigated in detail by several authors. This concept is discussed in detail within this chapter. An example from Turkey is also given how to design basic parameters of these machines and calculate production rates based of laboratory full-scale cutting tests.

It is also reported that it is possible to extend the limits of using continuous surface miners in harder formations by using mini discs in the cutting drums (Bilgin and Balci 1996).

References

Amar, P., Vemavarapu Mallika Sita Ramachandra, M., and Bahadur, S. K. 2013. Rock excavation using surface miners: An overview of some design and operational aspects. *International Journal of Mining Science and Technology,* 23(1):33–40.

Bilgin, N. and Balci, C. 1996. New excavation technologies in aggregate industry. *Proceedings of the* 1th *National Symposium on Aggregate Industry,* Turkish Chamber of Mines, Istanbul Branch, Turkey, pp. 207–225.

Bilgin, N., Copur, H., and Balci, C. 2014. *Mechanical Excavation in Mining and Civil Industries.* CRC Press/Taylor & Francis Group, Boca Raton, FL.

Dey, K. and Ghose, A. K. 2008. Predicting "cuttability" with surface miners: A rock mass classification approach. *Journal of Mines, Metals and Fuels,* 56(5–6):85–91.

Dey, K. and Ghose, A. K. 2011. Review of cuttability indices and a new rock mass classification approach for selection of surface miners. *Rock Mechanics and Rock Engineering,* 44:601–611.

Dutta, S. 2012. Application of surface miner in Indian coal mines. BSc. thesis, Department of Mining Engineering, National Institute of Technology Rourkela, Odisha, 56 p.

https://www.wirtgen-group.com/en/news-media/press-releases/wirtgen-surface-mining-maximizing-coal-recovery-by-minimizing-fines.106243.php, taken on May 2018, Wirtgen Group: News and Media.

Origliasso, C., Cardu, M., and Kecojevic, V. 2014. Surface miners: Evaluation of the production rate and cutting performance based on rock properties and specific energy. *Rock Mechanics and Rock Engineering*, 47(2):757–770.

Prakash Amar, P., Ramachandra, M.V.M.A., and Bahadur, S.K. 2013. Rock excavation using surface miners: An overview of some design and operational aspects. *International Journal of Mining Science and Technology*, 23:33–40.

Volk, H. J. 2016. Wirtgen drives the development of surface mining. "SYMPHOS," 3rd international symposium on innovation and technology in the phosphate industry, *Procedia Engineering*, 138:30–39.

Yazici, S., Acaroglu, O., Arapoglu, B., Bilgin, N., and Eskikaya, S. 1998. Investigation into of using continuous miners in an opencast coal mine. *Proceedings of the 8th Coal Congress of Turkey*, Zonguldak, Turkey, pp. 11–20.

Wirtgen, 2016. Wirtgen Surface Miners in Operation around the World, 24 p.

Index

Note: Page numbers in italic and bold refer to figures and tables respectively.